CW01661429

Python For Beginners

A crash course to learn Python Programming in 1 Week

Brady Ellison

Table of Contents

Introduction

Python is a general-purpose, high-level, interpreted programming language used for small and large scale coding projects. It has a comprehensive library of modules; it is multi-paradigm and emphasizes code readability.

Brief History of Python

Guido van Rossum was the lead developer in the Python development project conceived in the late 1980s. From December 1989 to July 2018, Python was the sole responsibility of Guido.

In January 2019, Guido, along with other core Python developers, selected a 5-person steering committee to help with Python's continued growth and development.

SETL was the most significant influence behind Python's development. Python was also the successor of the programming language ABC.

Python was first released to the public in 1991. Python's coding philosophy is based on providing an advanced developer and easier code readability. Its simple syntax requires a lot less code than most other programming languages.

One of the first Python releases was version 0.9.0, which was released in February 1991.

Version 2.0 was released in 2000. Version 2 ended in 2020, with the last update of the release being version 2.7.

In 2008 Python released version 3.0, which up until the release 3.5.x still supported Windows 7. Version 3.6.x and later no longer supports Windows 7 operating systems. The newer versions from 3.6.x and later also only support 64-bit installers for Windows operating systems.

Why Python?

Python grows from strength to strength with each new release. It was designed to be a scripted language meaning there is no need for complex compilers.

As a programming language, Python is neat, uncluttered, and its simple syntax structures make it easy to learn.

Python is free to distribute and use due to its OSI-approved open source license.

Python has a vast collaborating community of developers who continuously update and add-on modules to offer the language greater diversity.

Python has many third party modules – Python Package Index or PyPI. PyPI enables Python to interface with most of the other programming languages on the market.

Python is very easy to use and is a powerful language. It is one of the most popular and sought after trades in the current job market.

Chapter 1: Getting Started with Python

This chapter will set out the minimum standards a computer must meet to download and install Python. It will also cover the downloading and installation of Python.

Minimum Requirements for Installing Python

Before downloading or installing Python onto a Windows or Mac computer, you must ensure that the system meets the minimum requirements. The minimum standards ensure the integrity of the Python program operating performance.

Windows Requirements

- Windows 10
- 64-bit only
- 2 GB RAM is the bare minimum
- 8 GB RAM is the recommended requirement for optimum performance
- Hard Disk recommended is SSD with at least 2.5 GB free space
- Screen resolution should be at least 1024 x 768

Mac Requirements

- macOS 10.13 or newer

- 2 GB RAM is the bare minimum
- 8 GB RAM is the recommended requirement for optimum performance
- Hard Disk recommended is SSD with at least 2.5 GB free space
- Screen resolution should be at least 1024 x 768

Downloading Python

When you first download Python, you can get quite confused as there are two active versions of the software.

Python 2

Python version 2.7.18 is the latest version of this release. It is still actively used throughout the Python developers community. But it is not being developed or supported; it is the legacy version of Python.

Many programmers still use it because of the extensive library of modules, as this version has been around since 2000.

Python 3

Python version 3 is the latest version of the software and is currently being developed, supported, and maintained by Python. The newest release as of December 2020 was version 3.9.0, the version used throughout this tutorial.

Installing Python

To download Python, you will need to go to the Python website, which can be found by copying and pasting the following link into a web browser:

https://www.python.org/

The Python.org website will look similar to the image below:

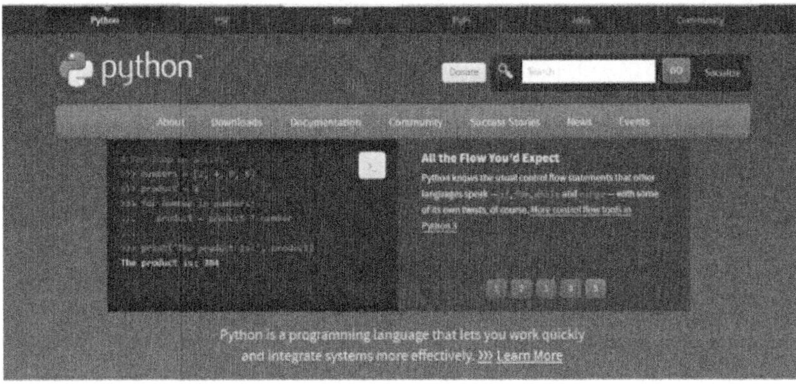

Choose the "Downloads" option from the menu bar to download the latest version of Python. At the time this training guide was written/printed, drop-down menus' right the latest versions of Python and the version used for this tutorial are as follows:

Windows Systems — Python 3.9.0

Mac Systems — Python 3.9.0

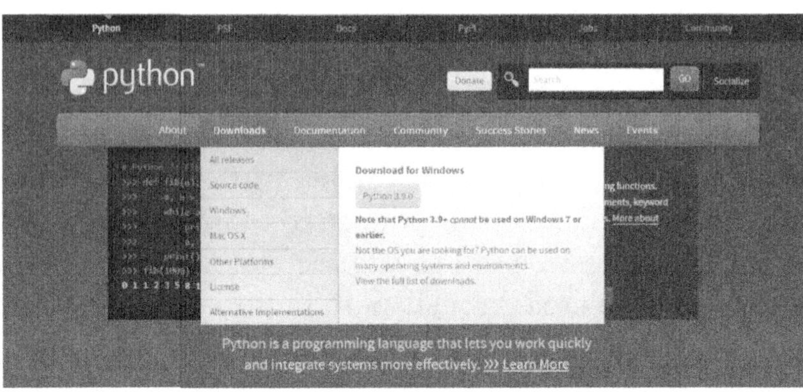

The "Download for Windows" option to the drop-down menus' right will change if you are using a Mac computer.

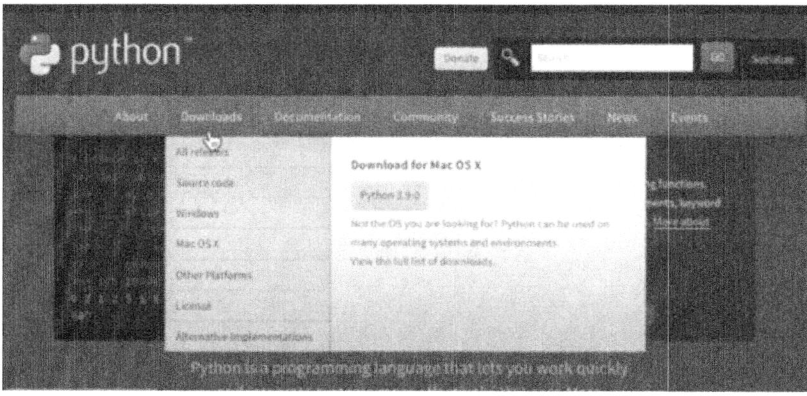

Installing Python 3 for Windows

When the "**Download**" option on the menu bar is clicked, the "**Download for Windows**" option appears as an option to the right of the drop-down menu.

Click on the gray "**Python 3.9.0**" button.

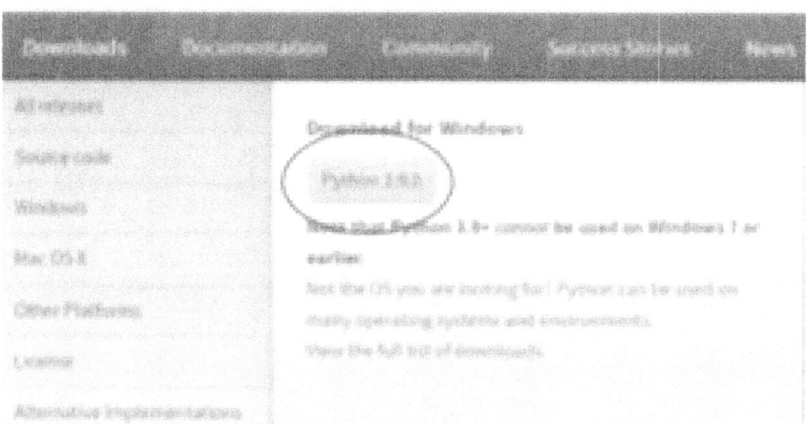

When you click the button, the following file gets downloaded to your computer's "**Downloads**" folder.

python-3.9.0-amd64.exe

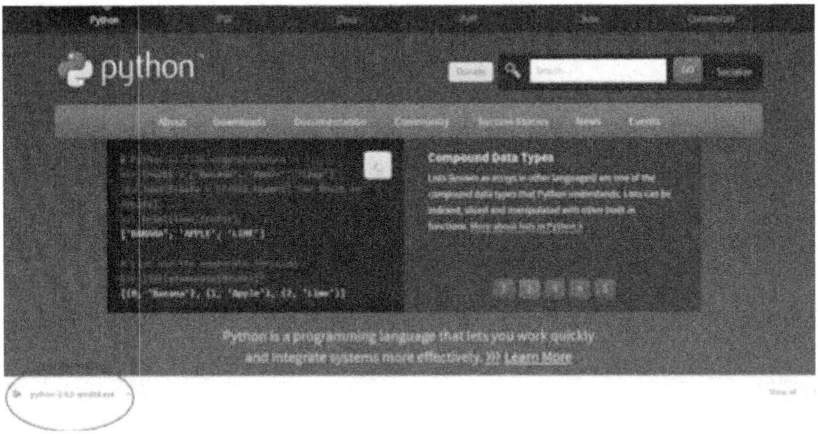

Go to the "**Downloads**" folder on your computer and double click on the "**python-3.9.0-amd64.exe**" file.

The first of the "**Install Python 3.9.0 (64-bit)**" screens will appear and look like the image below.

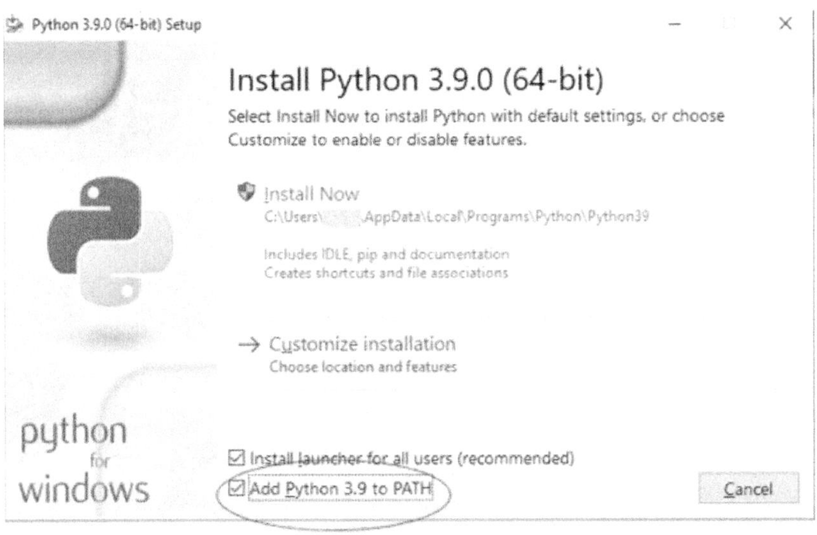

Click in the box next to the "**Add Python 3.9 to PATH**" option and ensure the box is checked before you click on the "**Install Now**" option.

Choose "**Yes**" to allow the program to make changes to the system.

The following screen shows the "**Setup Progress**" of the Python installation onto your computer.

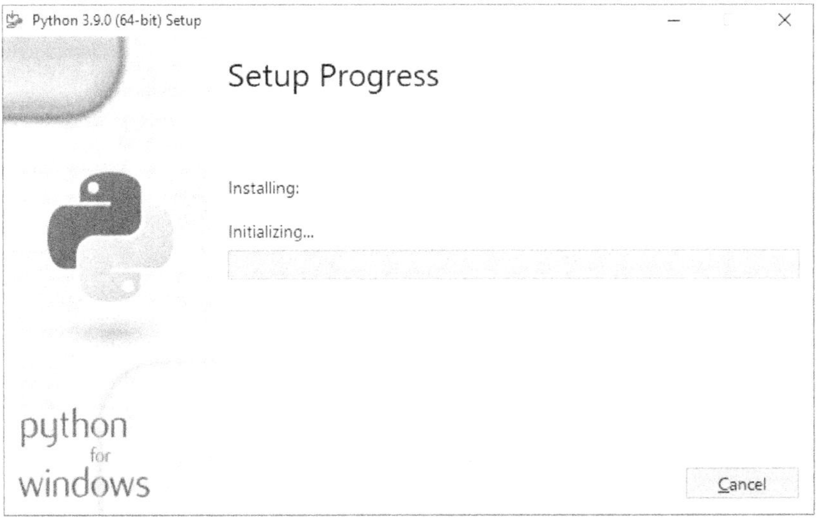

Once the download is complete; the following screen will appear:

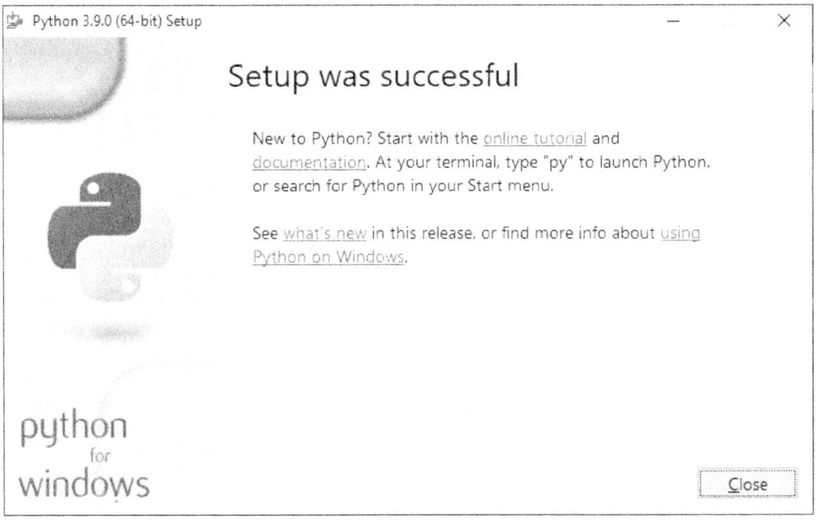

Click on the "**Close**" button to finish the installation.

It is always best practice to restart the machine once you have installed a new application.

Installing Python 3 for Mac

When the "**Download**" option on the menu bar is clicked, the "**Download for Mac OS X**" option appears as an option to the right of the drop-down menu.

Click on the gray "**Python 3.9.0**" button.

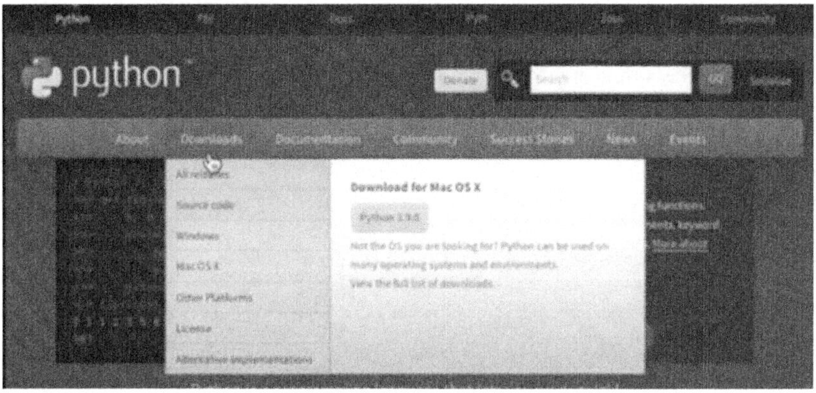

A pop-up box will appear once you have clicked on the grey button. The Pop-up box will ask you to confirm the download; the message will read:

> *Do you want to allow downloads on "www.python.org"?*

Choose "**Allow**"

The following file gets downloaded to the "**Downloads**" folders on the computer:

> *python-3.930-macosx10.pkg*

Go to the "**Downloads**" folder and double-click on the "python-3.930-macosx10.pkg" and the "**Python Installer**" screen will pop-up.

Click the "**Continue**" button at the bottom of the "**Introduction**" screen.

The next option will be the "**Read Me**" screen.

Click the “**Continue**” button at the bottom of the “**Read Me**” screen.

The next option will be the “**Licence**” screen.

You can “**Read Licence**” if you want.

Click the “**Agree**” button at the bottom of the “**Licence**” screen.

The “**Destination Select**” screen will appear.

Choose the default destination. Take note of the amount of space needed for the installation and make sure there is adequate space on the hard drive.

Click the “**Continue**” button at the bottom of the “**Destination Select**” screen.

The next option will be the “**Installation Type**” screen.

It is advisable to keep the default selections for the installation folder; for the sake of this tutorial, use the default folder.

Click the “**Install**” button at the bottom of the “**Installation Type**” screen.

The following **Python packages** get installed:

- Python Framework
- GUI Application
- UNIX command-line tools
- Python Documentation
- Shell profile updater
- Install or upgrade pip

At the “**Installer is trying to install new software**” screen, login to the Mac user account with administrative rights.

Click the “**Install Software**” button at the bottom of the “**Installer is trying to install new software**” screen.

The "**Installation**" screen is the next screen that will appear. This screen will prepare Python for installation and then start the installation.

The screen will have a progress bar that will indicate the progress of the Python installation.

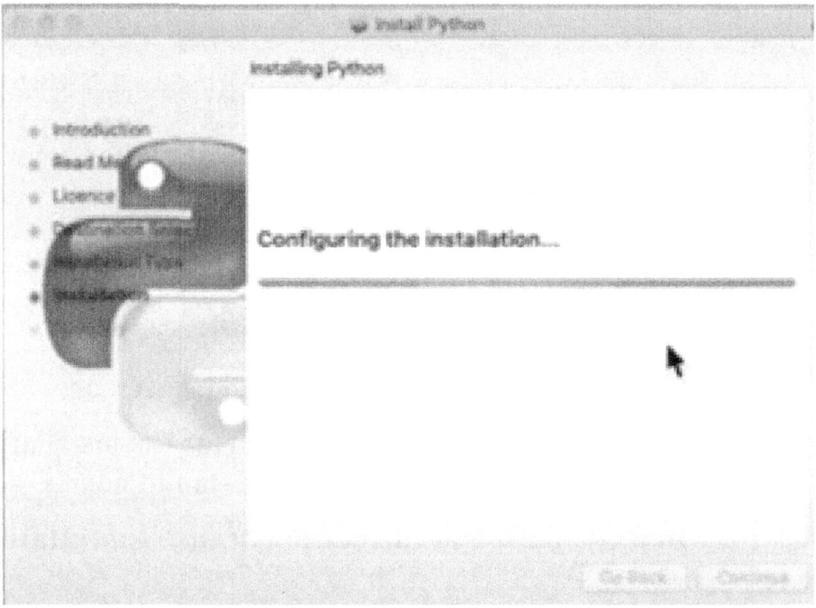

The "**Summary**" screen will display "**The installation was successful. The software was installed**" when the installation has been completed.

Click the "**Close**" button at the bottom of the "**Summary**" screen.

It is always best practice to restart the machine once you have installed a new application.

Using a Python Integrated Development Environment or Code Editor

Python is a programming language that you are going to be using to write code. To do this, you will need a program to write the code in and execute the written code.

If you were writing a document, you would need something like Microsoft Word or Google Docs to write your documents' content. You would need the program to save your document, edit the document, check for mistakes, share the paper, and so on.

If you think of Python code as content that needs to be written, you can think of a code editor similar to Microsoft Word or Google Docs. The type of code editor required depends on what kind of program a person is coding.

There are two basic types of code editing software packages available for coding, and they are:

- Code or Text Editing programs
- Integrated Development Environment (IDE) programs

What is an Integrated Development Environment (IDE)?

An Integrated Development Environment is software that incorporates a code editor, a debugger, build automation software, and many more built-in programming features.

If you look at the Microsoft Office suite, you will find that it incorporates a few programs housed under one software suite. An IDE is similar; only all the added features and programs are housed under one graphical user interface (GUI). Depending on the IDE, each has a host of functional and, at times, handy development tools. They are especially useful for machine learning applications.

What is a Code Editor?

A code editor is a basic editor in which a programmer can write their code. It does not come with all the fancy features or development tools of an IDE, but they still get the job done. Starting in a code editor is less confusing as a beginner can concentrate on coding and not get tied up in all the IDE features.

Code editors are similar to text editors such as a Notepad; only a code editor will have a few added features. These features include recognizing the loaded programming language, highlighting keywords, having a console for testing code functionality, highlighting coding errors, and more.

The Differences Between an IDE and a Code Editor

Features	IDE	Code Editor
Development tools	Comes packaged with development tools	Code Editor is a development tool used for writing and editing code.
Functionality	Most IDE's come standard with building, testing, code created as a single framework	Has built-in features to help simplify, edit, and manage code.
Key Features:	It has a GUI interface. Text editing Compiling Debugging Testing	Text editing Syntax Highlighting Console preview window Printing

	Syntax Highlighting Version Control Version Library Control More	

Python Coding Environment

There are many IDE's and Code Editors you could choose from. When deciding which one you want to use, you need to ensure that they have a few basic features. The following is a table of five standard features any good IDE or Code editor should have.

5 Standard Feature Requirements	Details of the Feature
Automatic Code Formatting	The code will automatically format itself for loops, statements, functions, etc., by automatically inserting space, comma's, and more.
Debugging	The software should provide for debugging with stepping. IDEs have more sophisticated debuggers than code editors do.
Code Execution Capable	IDEs and Code editors need to be able to execute code. Most will have some kind of Console or Terminal.

Save and Retrieve Code	Any file must be able to be saved. The saved file must be retrievable in the same state it was saved. The file must be editable.
Syntax Highlighting	A Syntax highlighter is used to highlight code errors and offer command line as well as syntax support.

Top 5 Python IDE's

Some of the top IDEs that work well with Python are as follows:

Idle

Python user level: *beginner, intermediate, and advanced Python programmers*

System Compatibility: Linux, macOS, and Windows

Download from: Idle comes bundled with Python.

Cost of the IDE: Open-source—free to download and use

Idle was released in 1998 along with Python by Guido Van Rossum and intended to be the Python programming IDE.

- Multi-window text editor
- Syntax highlighting
- Call stack visibility
- Debugger with stepping and persistent breakpoints
- Syntax highlighting, with smart indent, and auto-complete

- Console/Terminal
- It can be used to execute a single statement
- Can be used to create, edit, modify and delete Python scripts
- Idle needs a package manager to be installed onto a system running Python

PyCharm

Python user level: *beginner, intermediate, and advanced Python programmers*

System Compatibility: Linux, macOS, and Windows

Download from: JetBrains website from the following URL:

https://www.jetbrains.com/pycharm/

Cost of the IDE: Two versions of the IDE

1. Community version — Free, open-source software
2. Professional version— Paid for version (check the website for details)

PyCharm is developed and maintained by JetBrains. The interface is a good starting point for a beginner to learn even though it is aimed at the more advanced developer. It does not have a complex GUI and is easy to navigate.

- Multi-window text editor
- Syntax highlighting
- Call stack visibility
- Debugger with stepping and persistent breakpoints
- Error highlighting and suggestions

- Syntax highlighting, with smart indent, and auto-complete
- Console/Terminal
- Smart code navigation
- Can be used to create, edit, modify and delete Python scripts
- It is easy to download and install

PyDev/Eclipse

Python user level: *intermediate and advanced Python programmers*

System Compatibility: Linux, macOS, and Windows

Download from: eclipse website from the following URL:

https://www.eclipse.org/eclipseide/

Cost of the IDE: Open-source—free to download and use

PyDev/Eclipse was originally developed as a Java programming language IDE. PyDev converts Eclipse from a Javascript IDE into a Python IDE, bringing along all the powerful Java development tools.

- Comes with many complex integrated IDE features
- Multi-window text editor
- Syntax highlighting
- Code analysis
- Call stack visibility
- Debugger with stepping and persistent breakpoints
- Syntax highlighting, with smart indent, and auto-complete

- Console/Terminal
- It can be used to execute a single statement.
- It can be used to create, edit, modify, and delete Python scripts.
- It is easy to download but needs some prior knowledge of the system to correctly set up the software.
- Integrates smoothly with Django, which is used as Python's web development framework

Spyder

Python user level: *advanced Python programmers and data scientists*

System Compatibility: Linux, macOS, and Windows

Download from: Spyder website from the following URL:

https://www.spyder-ide.org/

Syder also comes bundled with the Anaconda Package Manager.

Cost of the IDE: Open-source—free to download and use

Syder is an IDE written in Python for a scientific environment and is designed for scientists, data analysis, and engineering. It is a good package for data exploration, deep inspection, and it has exceptional visualization capabilities.

- A multi-window text editor that comes with real-time analysis and introspection
- Syntax highlighting
- File explorer
- Call stack visibility

- Debugger with stepping and persistent breakpoints
- Error highlighting and suggestions
- Syntax highlighting, with smart indent, and auto-complete
- IPython console
- Profiler
- Static code analysis
- Smart code navigation
- Can be used to create, edit, modify and delete Python scripts
- History log
- It takes a more advanced user to install Sypder

Thonny

Python user level: beginners

System Compatibility: Linux, macOS, and Windows

Download from: Thonny website from the following URL:

https://thonny.org/

Cost of the IDE: Open-source—free to download and use

Thonny is one of the best IDEs for beginners as it was intended for the educational use of teaching Python to beginners. With the help of the Cybernetica AS and the RaspberryPi foundation, Thonny was developed in Estonia at the University of Tartu.

- This user interface of Thonny is straightforward.
- Multi-window text editor

- Syntax highlighting
- Call stack visibility
- Debugger with stepping and persistent breakpoints has been simplified to help the beginner better understand the process of debugging.
- Error highlighting and suggestions
- Syntax highlighting, with smart indent, and auto-complete
- Console/Terminal
- The editing window has line numbers making it easier for beginners to identify lines of code.
- It can be used to create, edit, modify, and delete Python scripts.
- It is easy to download and install, and it comes packaged with Python, so you only need one installation.
- Thonny has a handy user action log that logs all user interaction making it easier to track, trace, and analyze code.

Top 5 Python Code Editors

Some of the top Code Editors that work well with Python are as follows:

Atom

Python user level: *beginner, intermediate, and advanced Python programmers*

System Compatibility: Linux, macOS, and Windows

Download from: Atom website from the following URL:

https://atom.io/

Cost of the Code Editor: Open-source—free to download and use

Atom is a straightforward code editor that is easy to use, and it is highly customizable.

- Multi-window text editor
- Syntax highlighting
- smart auto-complete
- Debugger
- Error highlighting
- Syntax highlighting
- Smart indent
- Console/Terminal
- Can be used to create, edit, modify and delete Python scripts
- It is easy to download and install

GNU Emacs

Python user level: *beginner, intermediate, and advanced Python programmers*

System Compatibility: Linux, macOS, and Windows

Download from: GNU Emacs website from the following URL:

https://www.gnu.org/software/emacs/

Cost of the Code Editor: Open-source—free to download and use

GNU Emacs is one of the oldest code editors and was released in 1984. On the surface, it is a simple text editor that becomes a powerful, flexible, versatile, and easy to use code editor through plugins.

- Many customizable plug-ins
- Multi-window text editor
- Syntax highlighting
- Debugger
- Error highlighting
- Syntax highlighting
- Console/Terminal
- The editing window has line numbers
- Can be used to create, edit, modify and delete Python scripts
- It is easy to download and install

Notepad++

Python user level: *beginner, intermediate, and advanced Python programmers*

System Compatibility: Windows

Download from: Notepad-plus-plus website from the following URL:

https://notepad-plus-plus.org/downloads/

Cost of the Code Editor: Free to download and use

Notepad++ is a free code editor that supports a multitude of different programming languages. Notepad++ tends to use less

CPU power; it is written in C++ and is built to optimize execution time using STL and pure Win32API.

- Notepad ++ is a feature-rich code editor.
- Multi-window text editor
- Syntax highlighting
- Call stack visibility
- Debugger with stepping
- Error highlighting and suggestions
- Syntax highlighting
- Smart indent
- Auto-complete
- Console/Terminal
- The editing window has line numbers
- Can be used to create, edit, modify and delete Python scripts
- It is easy to download and install and configure for use with the Python programming language

Vim

Python user level: *beginner, intermediate, and advanced Python programmers*

System Compatibility: Linux, macOS, and Windows

Download from: Vim.org website from the following URL:

https://www.vim.org/download.php

Cost of the Code Editor: Open-source—free to download and use

Linux and macOS systems come pre-installed with the Vim text editor. However, it is available for download onto Windows systems. It can easily be configured to work with Python.

- Vim is a basic text editor.
- Multi-window
- Plugins can be used for:
    - ☐ Syntax highlighting
    - ☐ Code completion
    - ☐ Debugging
    - ☐ Error highlighting
- It is easy to download and install and configure to work with Python

Visual Studio Code

Python user level: *beginner, intermediate, and advanced Python programmers*

System Compatibility: Linux, macOS, and Windows

Download from: Visual Studio Code website from the following URL:

https://code.visualstudio.com/

Cost of the Code Editor: Open-source—free to download and use (the code editor)

Visual Code is part of the Visual Studio package but can be downloaded on its own as free, open-source software.

- Lightweight version of Visual Studio
- Feature-rich multi-window text editor
- Syntax highlighting

- Call stack visibility
- Debugger with stepping
- Error highlighting and suggestions
- Syntax highlighting
- Smart indent
- IntelliSense auto-complete
- Console/Terminal
- The editing window has line numbers.
- It can be used to create, edit, modify, and delete Python scripts.
- East to download and install

PyCharm

For the sake of this tutorial, you are going to be using one of the most popular Python IDE packages—**PyCharm**.

PyCharm can be download from a website called **JetBrains,** which can be found using the following URL:

https://www.jetbrains.com/pycharm/

Click on the "Download" button on either the top right-hand side of the screen (blue button) or in the middle of the screen (black button) to get to the "Download PyCharm" screen.

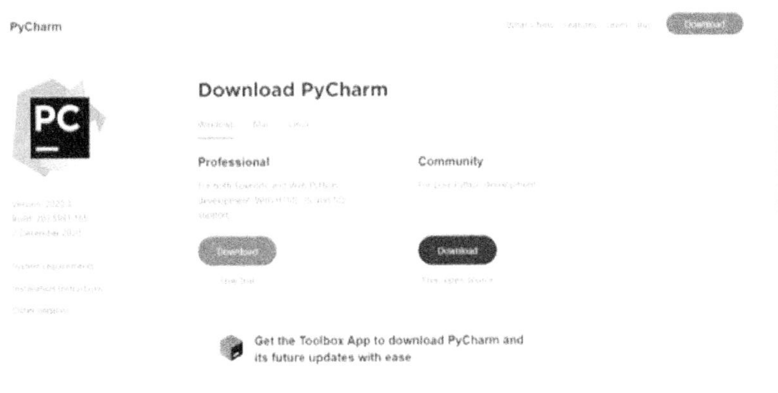

PyCharm for Windows

This first screen tag on the "Download PyCharm" screen will be the "Windows" download tag.

PyCharm Requirements for Windows

The following is currently the minimum systems requirements to run **PyCharm** on a **Windows** computer:

- Windows 8 or 10
- 64-bit only
- 2 GB RAM is the bare minimum
- 8 GB RAM is the recommended requirement for optimum performance
- Hard Disk recommended is SSD with at least 2.5 GB free space
- Screen resolution should be at least 1024 x 768
- Python version should be:
    - ☐ Python 2 = 2.7 or newer

- ☐ Python 3 = 3.5 or newer

Downloading and Installing PyCharm for Windows

The "Windows" option tag from the "**Download PyCharm**" screen must be selected.

The two versions of PyCharm are:

- **Professional** – This is the paid-for version
- **Community** – Free, open-source version

Click the "**Download**" button for the "**Community**" version of PyCharm.

The following file will be download into the "Downloads" folder on the computer:

pycharm-community-2020.3.exe

Go to the "Downloads" folder when the file has finished downloading.

Double-click on the "**pycharm-community-2020.3.exe**" file to start the installation of PyCharm.

A small pop-up box will appear to initiate the installation.

If the "**User Account Control**" screen appears, it will ask you:

> *Do you want to allow this app to make changes to your device?*

Click the "**Yes**" button at the bottom of the "**User Account Control**" screen.

The next pop-up box will be the "**PyCharm Community Edition Setup**" screen.

Click the "**Next**" button at the bottom of the "**Welcome to PyCharm Community Edition Setup**" screen.

It is advisable to keep the default directory allocated on the "**Choose Install Location**" screen.

Click the "**Next**" button at the bottom of the "**Choose Install Location**" screen.

The "Installation Options" screen needs the following options checked in the boxes next to them:

- Create Desktop Shortcut
    - ☐ 64-bit launcher
- Update PATH variable (restart needed)
    - ☐ Add launchers dir to the PATH

The next screen is the "**Choose Start Menu Folder,**" where you will find the PyCharm application. The default name for the folder is "**JetBrains**," this tutorial keeps the default folder name.

Click the "**Install**" button at the bottom of the "**Choose Install Location**" screen.

The next screen is the "**Installing**" screen, which will have a progress bar to indicate how far the PyCharm installation is to being finished.

When PyCharm has completed its installation, the "**Completing PyCharm Community Edition Setup**" screen will appear.

It is advisable to select the "**Reboot now**" option.

Click the "**Finish**" button at the bottom of the "**Completing PyCharm Community Edition Setup**" screen.

PyCharm for Mac

This second screen tag on the "Download PyCharm" screen will be the "Mac" download tag.

PyCharm Requirements for Mac

The following is currently the minimum systems requirements to run **PyCharm** on a **Mac** computer:

- macOS 10.13 or newer
- 2 GB RAM is the bare minimum
- 8 GB RAM is the recommended requirement for optimum performance
- Hard Disk recommended is SSD with at least 2.5 GB free space
- Screen resolution should be at least 1024 x 768
- Python version should be:
    - ☐ Python 2 = 2.7 or newer
    - ☐ Python 3 = 3.5 or newer

Downloading and Installing PyCharm for Mac

The "Mac" option tag from the "**Download PyCharm**" screen must be selected.

The two versions of PyCharm are:

- **Professional** – This is the paid-for version
- **Community** – Free, open-source version

Click the "**Download**" button for the "**Community**" version of PyCharm.

The following file will be download into the "Downloads" folder on the computer:

pycharm-community-2020.3.dmg

Go to the "Downloads" folder when the file has finished downloading.

Double-click on the "**pycharm-community-2020.3.dmg**" file.

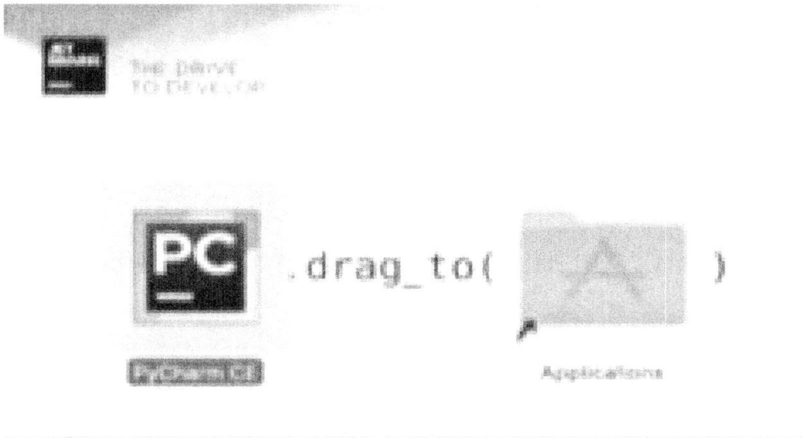

Drag the "**PyCharm CE**" picture on the left side and drop it onto the "**Applications**" folder on the right side.

When you first use PyCharm on a Mac, it will ask for a few settings to be configured. Until you are an advanced PyCharm or Python user, it is best to keep the **default** settings.

At the bottom of the "**Customize PyCharm**" screen, select the "**Skip Remaining and Set Defaults**" button.

Running Python

This tutorial uses **Windows 10**. There is not much difference in using PyCharm GUI on Mac or Windows. Mac users can still follow along with all the lessons set out in this tutorial.

Open the PyCharm program from the "Desktop" shortcut, the "Programs Menu," or search for it through the "Search" bar.

The first time you run PyCharm, the "**JetBrains Privacy Policy**" pop-up screen will appear.

Check the box next to the "**I confirm that I have read and accept the terms of this User Agreement**" box once you have read through the JetBrains Privacy Policy.

Click the "**Continue**" button for the "**JetBrains Privacy Policy**" screen.

At the "**Data Sharing**" screen, you can choose either **"Don't Send**" or "**Send Anonymous Statistics**." Be sure to read the screen thoroughly before allowing the program to "**Send Anonymous Statistics**" to JetBrains.

The tutorial uses the "**Don't Send**" button from the "**Data Sharing**" screen.

At the "**Welcome to PyCharm**" screen, there is the following option on the "**Main**" screen:

- New Project – To start a brand new Python programming project.
- Open – To open an existing Python programming project.
- Get from VCS – To get a document from a "**Version Control Library**."

Select "**New Project**" to get started on your first Python programming project.

Hello Me

Now that you have Python and PyCharm downloaded, you will write your first bit of Python code. Most programming courses start beginners off with the "Hello World" syntax. But, personalizing your first line of code is a lot more satisfying and makes you feel like your machine is greeting you and not the entire world.

Python is case sensitive, which means that any functions, variables, statements, etc., must be written in the correct case, or the program will throw out an error.

Choose "**New Project**" from the "**Welcome to PyCharm**" screen.

At the "**New Project**" screen, type in a name for the new project, call it "**HelloMe.**"

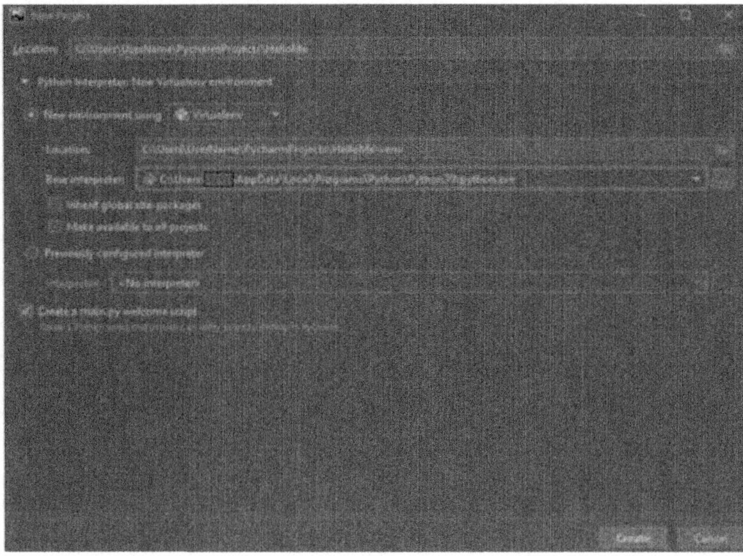

PyCharm will set up the project and the project's virtual environment.

At the “**Project**” screen, there will be “**Sample Code**” in the working area of the screen, which is the right-hand side of the screen.

Highlight any sample code and delete it.

Right-click on the “**HelloMe**” folder.

Choose “New”

Choose “Python File”

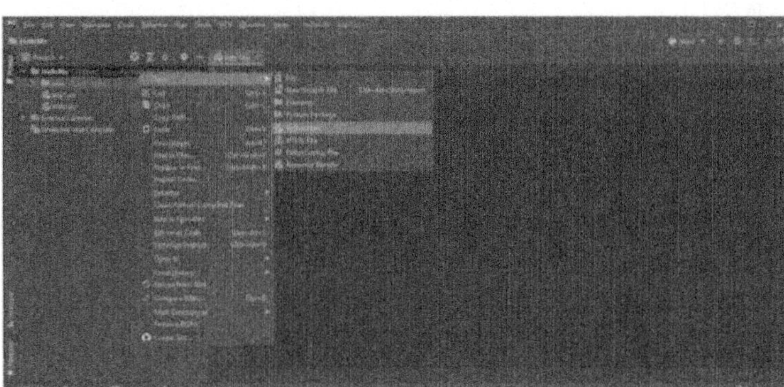

In the “New Python File” pop-up box, type “app1.”

Click in “**Line 1**” of the working area of the screen.

Type the following code:

print(“Hello Me”)

Press <**Enter**> to allow the cursor to progress to the next line in the workspace, or the program will throw out an error upon execution of the code.

You can run the code by right-clicking on the “**app1.py**” file from the screen’s left-hand side panel. Choose “**Run app1**” from the drop-down menu.

Or you can execute the program by going to the top menu bar, choosing the “**Run**” menu option. Choose “**Run**” from the drop-down menu. Choose “**app1.py**” from the pop-up file selection box.

The screen should like similar to the following:

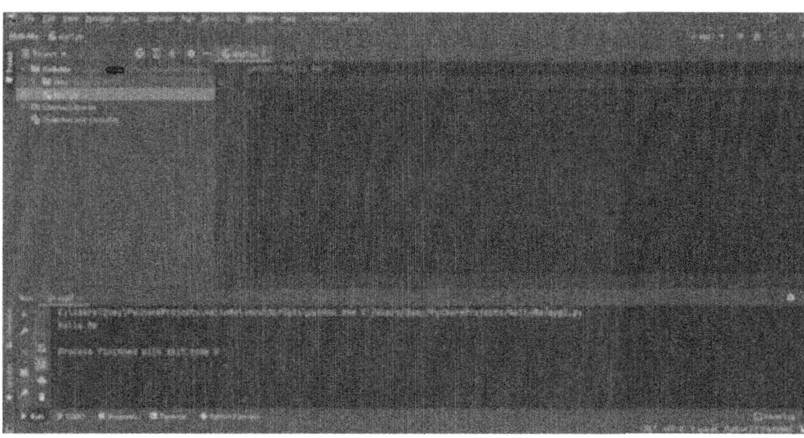

In the bottom “**Run**” panel, the text “**Hello Me**” should be printed out.

Well done, you have written your first bit of Python code.

Chapter 2: Getting Started with the Basics

When we write code into a code editor, we are essentially giving the computer a set of instructions to execute to deliver the required results.

Python is a popular programming language choice because it is an interpreted language.

An interpreted language means that the code does not need a compiler to convert it to machine code before being read by and executed by the system. Instead, Python is interpreted and run in a special run-time environment.

PyCharm Screen

Before getting started, take a look at the "**PyCharm Project**" screen.

At the top of the left-hand side of the screen is the **Menu Bar**.

The name of the "**Project Name**" and the "**Active Project File**" name appears to the **Menu Bar's** right.

To the right of the **Menu Bar** is the **Minimize** (—), **Full Screen** (▯), and **Exit (X)** icons.

Beneath the **Menu Bar** is the **Project Menu Bar**.

On the right of the **Project Menu Bar** is a drop-down menu selection. The **Quick Access** drop-down menu for accessing the current files in the project.

Next to the **Quick Access** drop-down menu are the "**Quick Access**" buttons. These buttons allow for quick access to **Run** (▷), **Debug (bug icon)**, and **Stop** (□) code from executing. The **Looking Glass** icon (⌕) is the **Search** facility.

Below that is the **Below Panel** that contains the **Project Files**, **Virtual Environments**, and **External Libraries**.

For this tutorial, you will mainly be working with the **Project Files**.

The Workspace window is where you will write the code. As you press enter on a new line and write code in it, it will be allocated a line number.

The bottom of the screen is called the "**Console**" or "**Terminal Window**" this is where any script execution, errors, and debug information will be shown.

Drawing Basic Shapes with the Print Function

To get a feel for how Python works, you will write some code that will instruct the machine to draw a simple shape using the **print()** function.

This exercise is very basic, but it will help you navigate the workspace and PyCharm screen.

Triangle

Create a new project and call it "**PrintShapes**"

Create a new Python file and call it "**Triangle**"

Starting in **Line 1**, type the following:

```
print("  /|")
print(" / |")
print("/__|")
```

Run the "**Triangle**" script.

The **Console** should display a triangle. It will look a lot like the older graphics looked.

Working with the Print() Function

The **print()** function is a built-in Python function, and it gets used quite a bit in Python. It is a good idea to learn a few essential tips about how the function works. It can also become a programmer's most significant ally when it comes to debugging tools.

As you have seen, the **Print** function is used to output the brackets' contents to the console.

The **Print** function looks as follows:

print('contents here')

Brackets must always follow the **Print** function, and inside the brackets, the content to be printed will be enclosed in either single or double-quotes.

Characters enclosed in quotes are known as a **string**.

There is no real difference in using double or single quotes in Python. Using either double or single quotes will affect the string's content if either version of the quotes is included in the string. For example:

print('They're over there')

The above example will return an **invalid syntax** error upon execution.

If single quotes are part of the string content, there are two ways to include them as string content without receiving an error.

Option 1

print("They're over there")

The example above uses double quotes to enclose the string; this will print the desired results without an error.

Likewise, you can use single quotes to enclose a string that needs to use double-quotes. For example:

print('"Hello World"')

When the above string is output to the console, it will print out as:

"Hello World"

Option 2

The second way to get around using a single quote in the string context is as follows:

print('They\'re over there")

Escape Characters

The **backslash (\)** in the above example is a special character, better known as an escape character. There are a lot of handy uses for this special character in the print() function. An escape character represents what is known as a "**whitespace**"; there are a few useful whitespace characters.

The following some useful escape characters with example exercises to try:

- \

  The backslash on its own tells the print() function to ignore the character directly after it.

  Example:

  print("\"Hello World\"")

Returns:

“Hello World”

Example:

print(“the directory is \\plums”)

Returns:

the directory is \plums

Note: The second example is how to print a backslash should the need arise.

- **\n**

  Backslash n inserts a new line between the content of the print() function string.

  Example:

  print(‘tulips\nroses’)

  Returns:

  tulips
  roses

- **\t**

  Backslash t inserts a tab space between the content of the print() function string.

  Example:

  print(“10\t20”)

  Returns:

  10 20

Chapter 3: Statements, Comments, Variables, and Index

This chapter will discuss what comments, variables, data types, and strings are and how to use them.

For this chapter, create a new project and call it **Ch3Projects**.

Statements

Statements are instructions that are understood by a Python interpreter and can be executed to carry out a specific function to return a required result.

Assignment statements are used to assign a value but do not produce a result.

```
a = 23
animals = ("Lion", "Tiger", "Bird")
```

The **print()** statement, for example, is a statement that produces a result.

```
print(a)
print(animals)
```

Python executes each line of code as you type it. For instance, you may not see the results, but as soon as the "animals" variable is assigned its values, Python has stored them in memory until they are ready to be used or reassigned.

A programming script will have more than one statement in it.

The **if()** statement is used to define a set condition for a statement where certain criteria must be met before the program can decide what to execute next.

Statements can be multi-line, in which case a line continuation would have to be inserted to mark the beginning and end of the statement. This can be done with the open and closed brackets.

```
number_game =     (1 + 2 + 3 +4 +
                  5 + 6 7 + 8 + 9)
```

A special character can also mark the end of each statement line:

```
number_game =     1 + 2 + 3 + 4 +\
                  5 + 6 7 + 8 + 9
```

The Importance of Indentation

In Python, to define a block of code, you need to indent it. Most of the interpreters automatically set the indentation levels as you write the code. If you do not get the indentations correct, the system will return an "IndentationError" when the code is executed.

Python enforces indenting code; this makes the code look neater and easier to read.

The body of a function, statements, loops, comments, docstrings, etc., usually contain code blocks that need to be indented in Python.

The general rule of thumb for indentation in Python is four white spaces. White spaces are used instead of tabs.

Example 1 of Indentation:

```
animal = ("Lion", "Whale", "Eagle",
          "Monkey", "Zebra")
```

```
print(animal)
```

Example 2 of Indentation:

```
def new_function(name):
  print("My name is" + name)
new_function("Mary")
```

Comments

Programmers annotate their work to reference what parts of their code are. Think about having to take over the coding on another programmer's job. Comments that detail and outline specific sections of the code will help you better understand the code. Thus, you would be able to take over or add to those sections quickly.

Without detailed comments, things could get quite messy and complicated. There are two types of comments a programmer can use in Python, **single-line comments** and **multi-line comments**.

To practice the examples in this section, create a new Python file under **Ch3Projects**, call it **ExComments**.

Single-line Comments

These comments comment on a specific line or section of code.

To insert a **single-line comment** into the Python script, you start the comment with a hash symbol (#). When the program comes across the hash (#) symbol, it will not process anything that comes after it.

In the example below, the single-line comment annotates what a particular line of code is.

Example:

Print(““Hello World””) #Demonstrates how to print out double quotes in a string.

Results:

“Hello World”

A single-line comment can start on any line to reference what a section of code is. In the example below, the comments are used to detail what the following lines of code do.

Example:

#Using the print() function with escape characters.

print(‘they\’re over there’)

print(‘tulips,\nroses’)

Results:

they’re over there

tulips
roses

The comments can start a section and be inserted after the code.

Example:

#Using the print() function with escape characters.

print(‘they\’re over there’) #Demonstrates the \ escape character

print(‘tulips, \nroses’) #Demonstrates the \n escape character

Results:

they’re over there

tulips
roses

Multi-line Comments

There are times when programmers need to offer a more detailed explanation of a section or piece of code. In this case, **multi-line comments** are the best type of comments to use. They are handy for writing a quick how-to guide, or an in-depth explanation of why they chose to write the code in that way, etc.

With the single-line comments, the hash (#) symbol only comments out the line's content. If you needed to write more than one line of comments using the hash (#) symbol, you would need to start each new line with the # symbol.

Example:

#This is an example of using a single-line comment to span multiple lines.

#Each new line of comment needs to start with a # symbol

When a very detailed guide is needed, starting each new line with a # can become problematic. Not only is it time-consuming, but it can also lead to a lot of unnecessary errors in the code.

However, using the # symbol for multi-line comments is the recommended standard for Python. But there is another more straightforward way: use quotation marks to define a **Text constant** of the content between the quotes.

Example

"""

This is a multi-line comment and can be used to span lines without having to use #

Using # on each line is the recommended Python way.

Using the quotations marks for comments has been known to cause errors on occasion.

""""

Variables

A variable is a temporary storage location or container that stores information known as a value in it. The variable will remain in use until the program is exited. Each time the program runs, the variable gets set to collect the corresponding values.

To follow the examples in this section, create a new Python file under **Ch3Projects**, call it **ExVariables**.

Working With Variables

The following code is an orders systems for a fruit shop:

print('Customer Z ordered apples')

When customer C orders oranges, the code has to be changed to:

print('Customer C ordered oranges')

The code is collecting data values on the following information:

- **Customer**
- **Item**

The code can be adjusted manually, but that would mean having to rewrite each line every time. Two lines of code are relatively easy to change, although even in those two lines, there is still room for human error. Imagine trying to manually adjust the values for a more extensive program with much more information to gather. The program would be a disaster and too time-consuming to update and troubleshoot.

Variables come in handy for the collection of values and data. Above are the two primary sources of information to be gathered for the program. The program would need the following containers to store all the information in; these are the program's **variables:**

- **Customer** – There will be more than one customer ordering fruit. You need their names to link up to what items they ordered.
- **Item** – This will be the item ordered, and unless there is only one product on sale, each customer's order will be different.

Creating, Assigning, and Implementing a Variable

To create a Python **variable**, you need to give the variable a name and then assign it a value.

Starting in Line 1 of the **ExVariables.py** tab (file), delete any existing code, and type the following:

customer = "Z"

item = "apples"

print("Customer " + customer + " ordered " + item)

Run **ExVarialble.py** to get the following results displayed in the console.

Customer Z ordered apples

When creating variables, the following rules need to apply:

- The variable name should reflect what it will be storing. In the example above the **customer, variable** stored the customer name, and the **item** variable stored the product ordered.
- When using more than one name for a variable, separate the name with an underscore; **customer_name**.

- Keep the variable name as short as possible.
- It is best practice to use all lowercase letters for variable names.
- The equal symbol (=) always follows the variable to link it to the variable's value.

It is important to take note of the spaces between the variable and equal signs.

It is also important to take note of any spaces within the value section of the variable.

Missing a space, inserting too many spaces, or inserting a space where there shouldn't be one, could cause syntax errors. Improper spacing could also fail to output the desired results.

A variable must be preceded with a plus (+) symbol to be inserted into a string.

print("Customer" + customer)

The above print function assumes that the code ends after the customer variable value is inserted.

If more code follows a variable value or another variable value needs to be added, the variable name needs to start and end with a plus (+) symbol.

print("Customer " + customer + " ordered " +item)

The above code will print out as follows because the plus (+) symbol at the end of the first variable tells the code there is more coding to follow.

Changing Variables

To change a variable in Python, you need to reassign the value of the named variable.

To change the value of the variable only requires the programmer to change the value of the assigned variable while the print function stays the same.

customer = "C"

Customer C ordered apples

Or you could change both variable values:

customer = "C"

item = "oranges"

To achieve the output results:

Customer C ordered oranges

The value(s) of a variable can be reassigned throughout a script.

Starting in Line 1 of the **ExVariables.py** tab (file), delete any existing code, and type the following:

customer = "Z"

item = "apples"

print("Customer " + customer + " ordered " +item)

customer = "C"

item = "oranges'

print("Customer " + customer + " ordered " +item)

Run **ExVarialble.py** to get the following results displayed in the console.

Customer Z ordered apples

Customer C ordered oranges

Index

To follow the examples in this section, create a new Python file under **Ch3Projects**, call it **ExIndex**.

A string or list in Python is a series of characters typed out in a systematic way to return the desired result. Each character in the string/list is assigned an index number, making the character easy to reference, lookup, extract, and manipulate.

Index

Each character in the string is given a number from zero (0) upwards. For instance, 1 2 3 4 5 to the end of the list/string.

For example, the following print function tells Python to print out the sentence Hello.

print("Hello")

The result will print:

Hello

Index: 12345

Python interprets the **H** as index number **0**, the **e** as index number **1**, the **l** as index number **2**, and so on.

Negative Index

Python is one of the few programming languages that accepts what is called a negative index.

A negative index works the same as a positive index (known as Index) does. But, instead of counting forward, the negative index counts backward from -1.

Clear any previous code from the **ExIndex** file, type the following in Line 1:

str1 = "sublime"

print(str1)

Result will print:

sublime

Negative Index interprets **e** as negative index **-1**, **m** as negative index **-2**, **i** as negative index **-3**, and so on.

The negative index counts the list of characters in a string from the last character to the string's first character—From right to left.

Index Function

Indexing can find where a specific character is located in a string or list.

Indexing is also handy for extracting desired information from a string.

Python has the **index()** built-in function to find and extract string values.

Clear any previous code from the **ExIndex** file, type the following in Line 1:

str1 = "sublime"

print(str1.index(e))

Result will print:

6

The above example returns the index position of the character **e** from the **str1** variable.

Clear any previous code from the **ExIndex** file, type the following in Line 1:

str1 = "sublime account"

print(str1.index(ount))

Result will print:

11

Whole words or parts of words can be used to find the word's index number, as demonstrated by the exercises above.

Chapter 4:
Data Types – Strings and Numbers

A data type is what categorizes and classifies the data items in a program.

Python has six different built-in data types:

The six data types in Python are:

- Strings
- Numbers
- Lists
- Sets
- Tuples
- Dictionaries

This chapter will explore the **Strings** and **Numbers** data types and how they work.

To follow the examples in this section, create a new Python project called **Ch4Projects**.

Strings

To follow the examples in this section, create a new Python file under **Ch4Projects**, call it **ExStrings**.

Strings are a **sequence** of **plain text/characters** enclosed within either **single** or **double-quotes**.

print("This sentence is a string enclosed in double-quotes")

print('This sentence is a string enclosed in single-quotes')

Strings **can be** used as a **value** of a **variable**

fname = "Harry"

lname = "Greenwood"

print("My name is " + fname + lname)

String literal

A string literal is the sequence of characters enclosed inside double or single quotation marks

Print("Hello")

"**Hello**" in the above code is the string literal.

String Variable

A string variable is a variable placed in the string.

name = "Peter"

print("Hello my name is " + name)

In the above code, the **name** is the string variable.

Strings are Immutable

Once a string has been declared, it cannot be changed or updated, known as immutable.

String Operations

Concatenation, **negative index**, **repetition**, and **slicing** are operations that can be performed on a string.

Concatenation

Concatenation operation is used to join two strings together. The **plus (+)** symbol joins the strings together.

Clear any previous code in the **ExStrings** file and type the following from Line 1:

str1 = "Sally Brown "

str2 = "has blue eyes"

print(str1 + str2)

Result will print:

Sally Brown has blue eyes

Repetition

Repetition instructs the program to print the same sequence of code for a set number of repetitions.

Clear any previous code from the **ExStrings** file and type the following in Line 1:

str1 = "laa dee dah "

*print(str1 * 3)*

Result will print:

laa dee dah laa dee dah laa dee dah

The plus (+) symbol adds two strings together, the multiplication (*) symbol multiplies the number of times a string needs to be repeated.

Replace

Parts of a string can be replaced by using the Replace function in Python.

Clear any previous code from the **ExStrings** file and type the following in Line 1:

str1 = "laa dee dah "

print(str1.replace("d", "t"))

Result will print:

laa tee dah

Change the print function to the following:

print(str.replace("dah", "row")

Result will print:

laa tee row

Slicing

Slicing is the technique Python programmers use to extract certain parts of code/characters from a string.

To understand and use the slicing operation, you need to understand the **Indexing** and **Negative Index,** as discussed in the sections above.

Clear any code in the **ExStrings** file and type the following in Line 1:

str1 = "Tone"

print(str1[1])

Result will print:

o

Change the print function to the following:

print(str[1:4])

Result will print:

one

Negative Indexes can be sliced as well.

Clear any code in the **ExStrings** file and type the following in Line 1:

str1 = "sublime"

print(str1)

Result will print:

sublime

Change the print function to reflect the following code:

print(str1[-4])

Result will print:

l

Change the print function to reflect the following code:

print(str1[-4:-1])

Result will print:

lim

Change the print function to reflect the following code:

print(str1[-4:])

Result will print:

lime

Numbers

To follow the examples in this section, create a new Python file under **Ch4Projects**, call it **ExNumbers**.

Numbers are one of the most commonly used data types in Python.

Numbers should not be enclosed in quotation marks unless they are being used in a normal string, for example:

print("Tim's birthday is January 7, 1989")

The above example identifies the numbers in the string as letter values instead of numeric ones. Thus, no arithmetic functions or operations can be performed on the numbers contained in the string.

Clear any code in the **ExNumbers** file and type the following in Line 1:

```
print(50)
```

Result will print:

```
50
```

Numbers can be calculated in Python as they would on a calculator or in a program such as Microsoft Excel, etc.

Working with Numbers

Python recognizes **positive numbers**, **negative numbers**, and numbers with **decimal places**.

Clear any code in the **ExNumbers** file and type the following in Line 1:

```
print(5 + 9.03)
```

Result will print:

```
14.03
```

Change the print function to reflect the following code:

```
print(-10 + 5)
```

Result will print:

```
-5
```

Parentheses can make more complex calculations in Python and classify the arithmetic operation's order to be calculated.

Calculation 1: There are ten classrooms; in each classroom, there are three teachers plus an extra seven teachers on standby. How many teachers are there altogether?

For the above scenario, the following calculation would work:

*print(10 * 3 + 7)*

Result will print:

37

But the above calculation would not work for Calculation 1. Order of arithmetic operations would need to be established. In Python, this is done by using parentheses.

Calculation 21: There are ten classrooms. In each classroom, there are three teachers and seven pupils. How many people are there in all ten classrooms?

*print(10 * (3 + 7))*

Result will print:

100

The above calculation will add what is in the brackets first. The total of that calculation will be multiplied by ten.

(3+7) = 10

*10 * 10 = 100*

Python follows the mathematical rules of **BODMAS,** which stands for:

> Brackets, Orders/Operations, Division, Multiplication, Addition, and Subtraction.

BODMAS is the order in which arithmetic operations work. For example:

B	(2 + 2)	Brackets get calculated first	2+2	= 4
O		Order of arithmetic operations		
D	(2 + 2) / 2	Division is calculated next	4/2	= 2

M	(2 + 2) / 2 * 3 2*3 = 6	Multiplication is calculated next
A	(2 + 2) / 2 * 3 + 1 6+1 = 7	Addition is calculated next
S	(2 +2) / 2 * 3 + 1 - 2 7-2 = 5	Subtraction is calculated next

Variables can contain **numbers** and **calculations**.

Clear any code in the **ExNumbers** file and type the following in Line 1:

numv1 = 9

print(numv1)

Result will print:

9

Clear any code in the **ExNumbers** file and type the following in Line 1:

numv1 = 9

numv2 = (9 + 1)

*print(numv * numv2)*

Result will print:

171

Numbers and Strings Working Together

There will be times when you will need to print a number alongside a string.

For example: "Tommy turned 9"

Clear any code in the **ExNumbers** file and type the following in Line 1:

numv1 = 9

print("Tommy turned " + numv1)

Result will print:

Traceback (most recent call last):

File "C:\Users\User\PycharmProjects\Ch3Projects\ExDataTypes.py", line 2, in <module>

print("Tommy turned " + numv1)

TypeError: can only concatenate str (not "int") to str

The code will not run and will return the above error. In order to use a number with a string, the number needs to be converted to a string.

A number can be converted to a string using the **str()** function in Python.

Change the print function to reflect the following code:

print("Tommy turned " + str(numv1))

Result will print:

Tommy turned 9

Built-in Python Mathematical Functions

Up to this point, you have worked with a few Python functions, mostly the print() function. A function is a piece of code that runs a command.

The **print()** function outputs what is contained in brackets to the console.

The **string()** function converts the brackets' value into a string value, and so on.

There are a few mathematical functions that can be used to perform more advanced mathematical operations.

abs()

The **abs()** function returns the absolute value of a number. Any negative signs that precede the value will be removed, as all values are perceived to be positive or zero (0)

Clear any code in the **ExNumbers** file and type the following in Line 1:

```
print(abs(-9))
```

Result will print:

```
9
```

The **abs()** function can be used with variables:

Clear any code in the **ExNumbers** file and type the following in Line 1:

```
numv1 = -9
print(abs(numv1))
```

Result will print:

```
9
```

max()

The **max()** function returns the largest value in a list of values.

Clear any code in the **ExNumbers** file and type the following in Line 1:

```
print(max(3, 6, 9))
```

Result will print:

```
9
```

The **max()** function can be used with variables:

Clear any code in the **ExNumbers** file and type the following in Line 1:

```
varn1 = 3
varn2 = 6
varn3 = 9
print(max(varn1, varn2, varn3))
```

Result will print:

```
9
```

min()

The **min()** function returns the smallest value in a list of values.

Clear any code in the **ExNumbers** file and type the following in Line 1:

```
print(min(3, 6, 9))
```

Result will print:

```
3
```

The **min()** function can be used with variables:

Clear any code in the **ExNumbers** file and type the following in Line 1:

```
varn1 = 3
varn2 = 6
varn3 = 9
print(min(varn1, varn2, varn3))
```

Result will print:

```
3
```

pow()

The **pow()** function is used to raise a value to the power of another value.

Clear any code in the **ExNumbers** file and type the following in Line 1:

```
print(pow(3, 3))
```

Result will print:

```
27
```

The **pow()** function can be used with variables:

Clear any code in the **ExNumbers** file and type the following in Line 1:

```
varn1 = 3

varn2 = 3

print(pow(varn1, varn2))
```

Result will print:

```
27
```

round()

The **round()** function is used to round a value to the nearest whole number.

Clear any code in the **ExNumbers** file and type the following in Line 1:

```
print(round(9.4))
```

Result will print:

```
9
```

The **round()** function will round any decimal value less than five down.

Clear any code in the **ExNumbers** file and type the following in Line 1:

```
print(round(9.5))
```

Result will print:

10

The **round()** function will round any decimal value from 5 up.

The **round()** function can be used with variables:

Clear any code in the **ExNumbers** file and type the following in Line 1:

```
varn1 = 9.6
print(round(varn1))
```

Result will print:

10

Types of Python Numbers

There are three defined classes of number data types in Python: complex, floating-point, and integer.

type()

The **type()** function can be used in a line of code to determine what class a number belongs to.

Clear any code in the **ExNumbers** file and type the following in Line 1:

```
a = 2
b = 9.5
c = 45h
print(type(a))
print(type(b))
print(type(c))
```

Result will print:

```
<class 'int'>
<class 'float'>
```

<class 'complex'>

Integers

An integer is a whole number. The whole number can be a positive or a negative and of infinitive length but it cannot contain a decimal or character such as 3x.

1 2 3 4 ... 89, and upward are integers

Complex Numbers

A complex number is expressed as a sum of a real number. They are written with a real part of the number and an imaginary part of the number. A complex number must follow the following format:

x + yj

Where x is the real part of the number, y is the number's imaginary part, followed by the letter j as the identifier.

(1 + 2j)

is a complex number

Floating Point Numbers

Floating point numbers are integers that have a decimal place. They can have up to fifteen decimal places.

23.567 is a floating point number

Exercise 1

In this exercise, you will use the **input()** function to prompt the user to enter their name and age using string variables.

To follow the examples in this section, create a new Python file under **Ch4Projects**, call it **Exercise1**.

Type the following in Line 1 of **Exercise1**

```
fname = input(“Enter your first name: ”)

lname = input(“Enter your last name: ”)

age = input(“Enter your age: ”)

print(“Hello ” + fname + “ ” + lname + “your age is ” + age)
```

Result will print:

Enter your first name: <type in Alice>

Enter your last name: <type in Brown>

Enter your age: <type in 22>

Hello Alice Brown your age is 22

Exercise 2

In this exercise you are going to use the **input()** function to prompt the user to enter some numbers and use the program as a simple calculator.

To follow the examples in this section, create a new Python file under **Ch4Projects**, call it **Exercise2**.

Type the following in Line 1 of **Exercise2**

```
a = input(“Enter number 1: ”)

b = input(“Enter number 2: ”)

answer = a + b

print(answer)
```

Result will print:

Enter number 1: <type in 5>

Enter number 2: <type in 10>

105

The answer above is not what you were expecting to get was it?

No, it was not! The answer you were supposed to get was 15.

Do you know the reason why Python output the incorrect answer?

Because the input() values are seen as a string, and string values cannot be calculated.

If you had to use multiply, division, or subtraction in the example above, an error would have been reported in the console.

When you started to explore numbers in this chapter, you learned how to convert a number into a string using the **str()** function.

To convert a string into a number, you can use either the:

- **int()** function – This will only allow the user to input a whole number. If a decimal number is put into the program will give an error.
- **float()** function – This will allow the user to input either an integer or decimal number.

Clear any code in the **Exercise2** file and type the following in Line 1:

```
a = input("Enter number 1: ")

b = input("Enter number 2: ")

answer = int(a) +int( b)

print(answer)
```

Result will print:

Enter number 1: <type in 5>

Enter number 2: <type in 10>

15

Clear any code in the **Exercise2** file and type the following in Line 1:

```
a = input(“Enter number 1: ”)

b = input(“Enter number 2: ”)

answer = float(a) +float( b)

print(answer)
```

Result will print:

Enter number 1: <type in 5.5>

Enter number 2: <type in 10.6>

16.1

Chapter 5: Data Types – List and Tuple

This chapter will explore the **List** and **Tuple** data types and how they work.

To follow the examples in this section, create a new Python project called **Ch5Projects**.

List

Python **lists** are data types that contain items in an ordered sequence. They are very versatile and commonly used throughout Python programming. The great thing about lists is that the items in a list do not have to be of the same type. Lists are also referred to as **arrays**.

Because lists are arrays of ordered values, they can be indexed and will allow duplicate items.

To follow the examples in this section, create a new Python file under **Ch5Projects**, call it **ExList**.

Creating a list is called "**declaring**" a list and is done by opening the list with **square brackets ([)** and ending the list with **square brackets (])** when the list is complete.

Items in the **list** must be separated by a **comma (,)**. Numeric items do not need quotation marks, but string items must be contained in quotation marks.

Items in a **list** can be **changed**, making **lists mutable**.

Clear any code in the **ExList** file and type the following in Line 1:

mylist = [2, 3, 'oranges', 'apples']

print(mylist)

Result will print:

[2, 3, oranges, apples]

List Index

A **lists index** is based on the item not the character, for instance in the following list:

myList = ["Elephant", "Tiger", "Monkey", "Bear", "Bird", "Rabbit"]

Index = Elephant = Index 0, Tiger = Index 1, Monkey = Index 2, Bear = Index 3, Bird = Index 4, Rabbit = Index 5

Negative Index = Elephant = Index -6, Tiger = Index -5, Monkey = Index -4, Bear = Index -3, Bird = Index -2, Rabbit = Index -1

A list can be **sliced** in the same way that a string can be sliced.

Clear any code in the **ExList** file and type the following in Line 1:

mylist = ["Elephant", "Tiger", "Monkey", "Bear", "Bird", "Rabbit"]

print(mylist[2])

Result will print:

Monkey

Clear any code in the **ExList** file and type the following in Line 1:

mylist = ["Elephant", "Tiger", "Monkey", "Bear", "Bird", "Rabbit"]

print(mylist[2:4])

Result will print:

['Monkey1, 'Bear']

The **[2:4]** gives the print() function a range of **list** items to print. It will print the items from index 2 up to, but not including, index 4.

Clear any code in the **ExList** file and type the following in Line 1:

mylist = ["Elephant", "Tiger", "Monkey", "Bear", "Bird", "Rabbit"]

print(mylist[2:])

Result will print:

['*Monkey', 'Bear', 'Bird', 'Rabbit'*]

Clear any code in the **ExList** file and type the following in Line 1:

mylist = ["Elephant", "Tiger", "Monkey", "Bear", "Bird", "Rabbit"]

print(mylist[2:4])

Result will print:

['Monkey1, 'Bear']

The **[2:]** gives the print() function a range of list items to print from index item 2 to the end of the list.

List items can be **modified** by their index number.

Clear any code in the **ExList** file and type the following in Line 1:

mylist = ["Elephant", "Tiger", "Monkey", "Bear", "Bird", "Rabbit"]

mylist[4] = "Wolf"

print(mylist[4])

Result will print:

Wolf

Functions Used With List Data Type

Many functions can be used with lists. This section will introduce the most commonly used **list** functions.

type()

The **type()** function will class the object as a data type '**list.**'

Clear any code in the **ExList** file and type the following in Line 1:

mylist = ["Elephant", "Tiger", "Monkey", "Bear", "Bird", "Rabbit"]

print(type(mylist))

Result will print:

<class 'list'>

list()

The **list()** function can be used to create a new list. Instead of the **square brackets ([])**, double **rounded brackets (( ))** are used.

Clear any code in the **ExList** file and type the following in Line 1:

mylist = list(("Elephant", "Tiger", "Monkey", "Bear", "Bird", "Rabbit"))

print(mylist)

Result will print:

['Elephant', 'Tiger', 'Monkey', 'Bear', 'Bird', 'Rabbit']

Check for an Item in a List

Use the **if** statement and **in** a keyword to determine whether an item is in a **list**.

Clear any code in the **ExList** file and type the following in Line 1:

```
mylist = ["Elephant", "Tiger", "Monkey", "Bear", "Bird", "Rabbit"]

if "Bird" in mylist :

print("Bird is in the list of animals")
```

Result will print:

Bird is in the list of animals

Modifying a Range of Items in a List

In a previous example, you modified one item in the **list**. A range of items can be changed/modified in the same instance in a **list**.

Clear any code in the **ExList** file and type the following in Line 1:

```
mylist = ["Elephant", "Tiger", "Monkey", "Bear", "Bird", "Rabbit"]

mylist[2:5] = ["Lion", "Rhino", "Wolf"]

print(mylist)
```

Result will print:

['Elephant', 'Tiger', 'Lion', 'Rhino', 'Wolf', 'Rabbit']

Modifying an Items in a List and Changing the Number of Items

A **list** can be modified and the numbers of items reduced.

Clear any code in the **ExList** file and type the following in Line 1:

mylist = ["Elephant", "Tiger", "Monkey", "Bear", "Bird", "Rabbit"]

mylist[1:6] = ["Lion", "Rhino"]

print(mylist)

Result will print:

['Elephant', 'Lion', 'Rhino']

insert()

Items can be inserted into a **list** using the **insert()** function.

Clear any code in the **ExList** file and type the following in Line 1:

mylist = ["Elephant", "Tiger", "Monkey", "Bear", "Bird", "Rabbit"]

mylist.insert (2, "Lion")

print(mylist)

Result will print:

['Elephant', 'Tiger', 'Lion', 'Monkey', 'Bear', 'Bird', 'Rabbit']

The new list item was inserted before the item indexed as 2.

append()

Items can be added to the end of a **list** using the **append()** function.

Clear any code in the **ExList** file and type the following in Line 1:

mylist = ["Elephant", "Tiger", "Monkey", "Bear", "Bird", "Rabbit"]

mylist.append ("Lion")

print(mylist)

Result will print:

['Elephant', 'Tiger', 'Monkey', 'Bear', 'Bird', 'Rabbit', 'Lion']

The new list item was appended to the end of the list.

extend()

Items from one **list** can be appended to another **list** using the **extend()** function.

Clear any code in the **ExList** file and type the following in Line 1:

mylist = ["Elephant", "Tiger", "Monkey", "Bear", "Bird", "Rabbit"]

mylist2 = ["Eagle", "Lion", "Wolf", "Owl"]

mylist.extend(mylist2)

print(mylist)

Result will print:

['Elephant', 'Tiger', 'Monkey', 'Bear', 'Bird', 'Rabbit', 'Eagle', 'Lion', 'Wolf', 'Owl']

The two lists are joined together.

remove()

Items can be removed from a **list** using the **remove()** function.

Clear any code in the **ExList** file and type the following in Line 1:

mylist = ["Elephant", "Tiger", "Monkey", "Bear", "Bird", "Rabbit"]

mylist.remove("Bird")

print(mylist)

Result will print:

['Elephant', 'Tiger', 'Monkey', 'Bear', 'Rabbit']

The bird will have been removed from the list of items.

pop()

Specific items can be removed from the **list** using the index number; this is done with the **pop()** function.

Clear any code in the **ExList** file and type the following in Line 1:

mylist = ["Elephant", "Tiger", "Monkey", "Bear", "Bird", "Rabbit"]

mylist.pop(4)

print(mylist)

Result will print:

['Elephant', 'Tiger', 'Monkey', 'Bear', 'Rabbit']

The bird will have been removed from the list of items.

Keyword – del

The keyword **del** can remove a specified item from the list using items indexed number.

Clear any code in the **ExList** file and type the following in Line 1:

mylist = ["Elephant", "Tiger", "Monkey", "Bear", "Bird", "Rabbit"]

del mylist[4]

print(mylist)

Result will print:

['Elephant', 'Tiger', 'Monkey', 'Bear', 'Rabbit']

The bird will have been removed from the list of items.

The **del** keyword can also be used to delete an entire list of items.

Clear any code in the **ExList** file and type the following in Line 1:

mylist = ["Elephant", "Tiger", "Monkey", "Bear", "Bird", "Rabbit"]

del mylist

The entire list will be erased.

clear()

The **clear()** function can be used to clear items from a list.

Clear any code in the **ExList** file and type the following in Line 1:

mylist = ["Elephant", "Tiger", "Monkey", "Bear", "Bird", "Rabbit"]

mylist.clear()

print(mylist)

Result will print:

[]

The **print()** function will return an empty list.

sort()

The **sort()** function can be used to sort items in a **list** in ascending and descending order. The **sort()** function can be used on both string and numeric values in a **list**.

Clear any code in the **ExList** file and type the following in Line 1:

mylist = ["c", "e", "b", "d", "a", "f"]

mylist.sort()

print(mylist)

Result will print:

['a', 'b', 'c', 'd', 'e', 'f']

The **sort()** function will sort the alphanumeric (string) items in the list alphabetically in ascending order.

Clear any code in the **ExList** file and type the following in Line 1:

mylist = [10, 4, 6, 8, 2]

mylist.sort()

print(mylist)

Result will print:

[2, 4, 6, 8, 10]

The **sort()** function will sort the numeric items in the list in ascending order.

The **sort()** function will sort both alphanumeric (string) items and numeric items in a list in descending order using **sort(reverse = True)**.

Clear any code in the **ExList** file and type the following in Line 1:

```
mylist = ["c", "e", "b", "d", "a", "f"]
mylist1 = [10, 4, 6, 8, 2]
mylist.sort(reverse = True)
mylist1.sort(reverse = True)
print(mylist)
print(mylist1)
```

Result will print:

```
['f', 'e', 'd', 'c', 'b', 'a']
[10, 8, 6, 4, 2]
```

reverse()

The **revers()** function will reverse the order of the items in the list.

Clear any code in the **ExList** file and type the following in Line 1:

```
mylist = ["ant", "wasp", "bee"]
mylist.reverse()
print(mylist)
```

Result will print:

```
['bee', 'wasp', 'ant']
```

2D Lists

A 2D list is used to form a grid structure in Python, meaning that you create a list structure that has numerous columns and rows.

Clear any code in the **ExList** file and type the following in Line 1:

```
mylist = [
```

```
    [10, 20, 30, 40],
    [50, 60, 70, 80],
    [0]
    ]
print(mylist)
```

Result will print:

[[10, 20, 30, 40], [50, 60, 70, 80], [0]]

2D List Index

Index of a **2D list** works with rows starting with row 0, up. Each row will index the items in the list as per normal list item index.

Index Row 0 *[10, 20, 30, 40],*

Index Row 1 *[50, 60, 70, 80],*

Index Row 2 *[0]*

Items in a **2D list** can be modified, accessed, and deleted the same as you would for a normal list.

Clear any code in the **ExList** file and type the following in Line 1:

```
mylist = [
    [10, 20, 30, 40],
    [50, 60, 70, 80],
    [0]
    ]
print(mylist[0][2])
```

Result will print:

30

Tuple

A tuple is similar to a list, but unlike lists, a tuple is immutable— Once a tuple is defined, it cannot be changed. One of the main uses of a tuple in a program is to write-protect data. Because a tuple cannot dynamically change, it is a lot faster than a list.

Because tuples contain ordered values, they can be indexed and will allow duplicate items.

A tuple is declared within **rounded brackets ().**

To follow the examples in this section, create a new Python file under **Ch5Projects**, call it **ExTuple**.

Clear any code in the **ExTuple** file and type the following in Line 1:

```
tup1 = (9, 10, "Dove", "Crab", 8)
print(tup1)
```

Result will print:

```
(9, 10, 'Dove', 'Crab', 8)
```

Data can be extracted from a tuple, but it cannot be manipulated or modified within the tuple.

Clear any code in the **ExTuple** file and type the following in Line 1:

```
tup1 = (9, 10, "Dove", "Crab", 8)
print(tup1[2])
```

Result will print:

```
Dove
```

Chapter 6:
Data Types – Set and Dictionary

This chapter will explore the **Set** and **Dictionary** data types and how they work.

To follow the examples in this section, create a new Python project called **Ch6Projects**.

Set

A **set** is a data-type much like tuples and lists; it is used to store a collection of values/data. Unlike tuples and lists, the items in a set are unordered and unindexed.

Because a set is an unordered list of values, it cannot be indexed and will not allow duplicate items in a **set**. You can add values to a defined set, but you cannot change defined values.

Sets are defined within **curly brackets {}**.

To follow the examples in this section, create a new Python file under **Ch6Projects**, call it **ExSet**.

Clear any code in the **ExSet** file and type the following in Line 1:

setno1 = {"pink"", "blue", "green", "yellow")

print(setno1)

Result will print:

('pink', 'blue', 'yellow')

If you tried to change or modify an item in a set, Python would give an error.

Dictionary

Dictionary's store values using a **key identifier**. They are one of the most flexible built-in data types in Python to store significant data quantities.

To define **dictionary** items, **curly brackets {}** are used. Because a key is used to describe the dictionary item, you would have to know the item's key to retrieve it. A dictionary item cannot be recovered without using the item's key.

Dictionary items do not allow duplicates as they store unordered sequences of mutable data values (the values can be changed).

To follow the examples in this section, create a new Python file under **Ch6Projects**, call it **ExDictionary**.

Clear any code in the **ExDictionary** file and type the following in Line 1:

dict1 = {1: "apples", 2: 1, 3: "dogs"}

print(dict1)

Result will print:

{1: 'apples', 2: 1.3, 3: 'dogs'}

The **keys** in the above example are **1:**, **2:**, and **3:** these are the **keys** a programmer will use to retrieve or manipulate data values in a **dictionary**.

Clear any code in the **ExDictionary** file and type the following in Line 1:

dict1 = {1: "apples", 2: 1, 3: "dogs"}

print(dict1[3])

Result will print:

dogs

The **key** can be used to update or modify data values in a **dictionary**.

Clear any code in the **ExDictionary** file and type the following in Line 1:

dict1 = {1: "apples", 2: 1, 3: "dogs"}

dict1[2] = 5

print(dict1)

Result will print:

{1: 'apples', 2: 5, 3: 'dogs'}

The **key** can be used to add data values in a **dictionary**.

Clear any code in the **ExDictionary** file and type the following in Line 1:

dict1 = {1: "apples", 2: 1, 3: "dogs"}

dict1[4] = 5

print(dict1)

Result will print:

{1: 'apples', 2: 1.3, 3: 'dogs', 4: 5}

A **dictionary** is useful when you need to define a table of items for a computer shop and need the make, model, and year the item was built.

Clear any code in the **ExDictionary** file and type the following in Line 1:

compmods = {

```
    'make' : 'Dell',
    'model' : 'E5640',
    'year' : 2010
}
print(compmods)
```

Result will print:

```
{'make': 'Dell', 'model': 'E5640', 'year': 2010}
```

Chapter 7: Operators

This chapter will explore the various operators in Python and how they work.

An operator is used to carry out **arithmetic**, **logical**, and **comparison** operations on an operand.

Operators are used in the form of a special symbol such as +, *, >, etc.

To follow the examples in this section, create a new Python project called **Ch7Projects**.

Arithmetic Operators

An **arithmetic operator** is used to perform mathematical operations between a set of numbers.

These operators are the same and work much the same as they do with regular maths.

To follow the examples in this section, create a new Python file under **Ch7Projects**, call it **ExAO.**

Clear any code in the **ExAO** file and type each of the following examples listed in the table below in Line 1.

Python has **seven** basic **arithmetic operators**.

Arithmetic Operator	Description	Example	Print Result
+	Addition – Adds a set of numbers.	*print(10 + 20)*	*30*
-	Subtraction – Subtracts a set of numbers.	*print(30 - 10)*	*20*
/	Division – Divides a set of numbers.	*print(30 / 10)*	*3*
//	Floor Division – Rounds divided results down to the nearest whole number.	*print(30 // 8)*	*3*
*	Multiplication – Multiplies a set of numbers.	*print(30 * 8)*	*240*
**	Exponentiation – Exponentiates a number by the given multiple. 30**3 is the same as: 30*30*30	*print(30 ** 3)*	*27000*
%	Modulo operation – This operator returns the leftover number when dividing a set of numbers. Suppose you think of it as a pie with nine slices to be divided evenly by four people. The Modulo operator is going to be one.	*print(5 % 2)* *print(10 % 5)*	*1* *0*

	9/4 = 2 with 1 left over		

Comparison Operators

A **comparison operator** is used to decide on a relationship between two values. Hence the reason why a **comparison operator** is referred to as a **relational operator**.

These operators are the same and work much the same as they do with regular maths.

For example:

```
a = 10

b = 10

print(b==a)
```

The above example determines whether **b** is **equal** to **a**. If the **value** in **b** is the **same** as the **value** in **a,** the program will return a **True** result. Otherwise, it will return a **False** result.

To follow the examples in this section, create a new Python file under **Ch7Projects**, call it **ExCO.**

Clear any code in the **ExCO** file and type each of the following examples listed in the table below in Line 1.

Python has **six** basic **comparison operators**.

Comparison Operator	Description	Example	Print Result
!=	Not Equal To If value a is not equal to value b, the condition will be True. If value a is equal to value b, the condition	*a = 10* *b = 15* *print(a != b)*	*True*

	will be False.		
==	Equal To If value a is equal to value b, the condition will be True. If value a is not equal to value b, the condition will be False.	*a = 10* *b = 10* *print(a == b)*	*True*
>	Greater Than If value a is greater than value b, the condition will be True. If value a is less than value b, the condition will be False.	*a = 15* *b = 10* *print(a > b)*	*True*
>=	Greater Than or Equal To If value a is greater than or equal to value b, the condition will be True. If value a is less than value b, the condition will be False.	*a = 15* *b = 10* *print(a >= b)*	*True*
<	Less Than If value a is less than value b, the condition will be True. If value a is greater than value b, the condition will be False.	*a = 10* *b = 15* *print(a < b)*	*True*
<=	Less Than or Equal To	*a = 10*	*True*

	If value a is less than or equal to value b, the condition will be True. If value a is greater than value b, the condition will be False.	*b = 10* *print(a <= b)*	

Logical Operators

A **logical operator** is used to perform logical and, or, and not operations on conditional statements. A logical operator is commonly used with if - then - else statements.

For example:

```
a = 10

b = 20

if a != 0 and b>= a:

  print("This condition is True")
```

These operators are the same and work much the same as they do with standard conditional statements.

To follow the examples in this section, create a new Python file under **Ch7Projects**, call it **ExLO.**

Clear any code in the **ExLO** file and type each of the following examples listed in the table below in Line 1.

Python has **three** basic **logical operators**.

Logical Operator	Description	Example	Print Result
and	The logical "and" operator will return True if both values of the comparison	*a = 8* *b = 10* *if a< 10 and b =*	*True*

	statements are True (both values meet the given condition).	*10:* *print("True")*	
not	The logical "not" operator will return True if the value of the comparison statement is False (the value does not meet the given condition).	*a = 10* *if not (a > 10):* *print("a <= 10")* *else:* *print("a is > 10")*	*a <= 10*
or	The logical "or" operator will return True if one or more conditions of the comparison statement are true (one or more of the values meet the given condition).	*a = 20* *b = 25* *if a >= 10 or b <= 25:* *print("True")* *else:* *print("False")*	*True*

Assignment Operators

An **assignment operator** is used to assign a value(s) to a variable(s).

For example:

a = 10

There are a few compound assignment operators that assign a value and an operator to a variable.

For example:

a = 10

a +=5

print(a)

Result will print:

15

To follow the examples in this section, create a new Python file under **Ch7Projects**, call it **ExAssO.**

Clear any code in the **ExAssO** file and type each of the following examples listed in the table below in Line 1.

Python has **eight** basic **assignment operators**.

Assignment Operator	**Description**	**Example**	**Print Result**
=	Assigns a variable and gives it a value(s). The value assigned to the left of the "=" is the variable operand. The value assigned to the right of the "=" is the value operand.	*a = 10* *b = "Ten"* *print(a)* *print(b)*	*10* *Ten*
+=	Adds the left variable to the right value and assigns the results to the variable	*a = 10* *a += 5* *print(a)* This is	*15*

	operand to the left of the "+=" operator. The value assigned to the left of the "+=" is the variable operand. The operator "+" is the operation to be performed on the value operand. The value assigned to the right of the "+=" is the value operand.	equivalent to: *a = 10* *a = a + 5* *print(a)* Or: *a = 10* *b = 5* *print(a + b)*	
-=	Subtracts the left variable from the right value and assigns the results to the variable operand to the left of the "+=" operator. The value assigned to the left of the "-=" is the variable	*a = 10* *a -= 5* *print(a)* This is equivalent to: *a = 10* *a = a - 5* *print(a)* Or:	*5*

	operand. The operator "-" is the operation to be performed on the value operand. The value assigned to the right of the "-=" is the value operand.	*a = 10* *b = 5* *print(a - b)*	
*=	Multiplies the left variable to the right value and assigns the results to the variable operand to the left of the "*=" operator. The value assigned to the left of the "*=" is the variable operand. The operator "*" is the operation to be performed on the value operand. The value	*a = 10* *a *= 5* *print(a)* This is equivalent to: *a = 10* *a = a * 5* *print(a)* Or: *a = 10* *b = 5* *print(a * b)*	*50*

	assigned to the right of the "*=" is the value operand.		
/=	Divides the left variable from the right value and assigns the results to the variable operand to the left of the "/=" operator. The value assigned to the left of the "/=" is the variable operand. The operator "/" is the operation to be performed on the value operand. The value assigned to the right of the "/=" is the value operand.	*a = 10* *a /= 5* *print(a)* This is equivalent to: *a = 10* *a = a / 5* *print(a)* Or: *a = 10* *b = 5* *print(a / b)*	*2*
%=	Returns the modulus of the left variable and	*a = 16* *a %= 5*	*1*

	the right value. The results are assigned to the variable operand to the left of the "+=" operator. The value assigned to the left of the "%=" is the variable operand. The operator "%" is the operation to be performed on the value operand. The value assigned to the right of the "%=" is the value operand.	*print(a)* This is equivalent to: *a = 16* *a = a % 5* *print(a)* Or: *a = 16* *b = 5* *print(a % b)*	
**=	Calculates the exponential value of the left variable by the right value and assigns the results to the variable operand to the left of the "**=" operator. The value	*a = 5* *a **= 3* *print(a)* This is equivalent to: *a = 5* *a = a ** 3*	*125*

	assigned to the left of the "**=" is the variable operand. The operator "**" is the operation to be performed on the value operand. The value assigned to the right of the "**=" is the value operand.	*print(a)* Or: *a = 5* *b = 3* *print(a ** b)*	
//=	Calculates the floor division of the left variable by the right value and assigns the results to the variable operand to the left of the "//=" operator. The value assigned to the left of the "//=" is the variable operand. The operator "//" is the operation to be performed on	*a = 9* *a //= 3* *print(a)* This is equivalent to: *a = 9* *a = a // 3* *print(a)* Or: *a = 9* *b = 3* *print(a // b)*	*3*

	the value operand. The value assigned to the right of the "//=" is the value operand.		

Bitwise Operators

A **bitwise operator** is used for integer bitwise calculations. The integer is first converted into a binary number upon which binary comparison and operations are performed bit-by-bit.

bin()

To find out the binary value of an integer, the bin() function can be used.

To find the binary value of the integer 70:

print(bin(70))

Result will print:

0b1000110

To follow the examples in this section, create a new Python file under **Ch7Projects**, call it **BitO.**

Clear any code in the **BitO** file and type each of the following examples listed in the table below in Line 1.

Python Bitwise Operators.

Bitwise Operator	Description	Example	Print Result

<table>
<tr><td>&</td><td>The ampersand translates to “and” operator.
If each</td><td>a = 5
b = 9
print(x & y)
Binary 5 = 00101
Binary 9 = 001001</td><td>1</td></tr>
<tr><td>|</td><td>The logical “not” operator will return True if the value of the comparison statement is False (the value does not meet the given condition).</td><td>a = 10
if not (a > 10):
  print("a <= 10")
else:
  print("a is > 10")</td><td>a <= 10</td></tr>
<tr><td>^</td><td>The logical “or” operator will return True if one or more conditions of the comparison statement are true (one or more of the values meet the given condition).</td><td>a = 20
b = 25
if a >= 10 or b <= 25:
  print("True")
else:
  print("False")</td><td>True</td></tr>
</table>

Identity Operators

An **identity operator** is used to compare objects that occupy the same memory location.

To follow the examples in this section, create a new Python file under **Ch7Projects**, call it **ExIDO.**

Clear any code in the **ExIDO** file and type each of the following examples listed in the table below in Line 1.

Python has **two** basic **identity operators**.

Identity Operator	Description	Example	Print Result
is	The "is" identity operator evaluates an object to establish if the objects are the same. If the objects are the same, it will return a True value. If the objects are not the same, it will return a False value.	Exercise 1: *a = 8* *b = 10* *print(a is b)* Exercise 2: *a = 8* *b = 10* *c = a* *print(a is c)*	*False* *True*
is not	The "is not" identity operator evaluates an object to establish that the objects are not the same. If the objects are the same, it will return a False value. If the objects are not the same, it will	Exercise 1: *a = 8* *b = 10* *print(a is not b)* Exercise 2: *a = 8* *b = 10* *c = a* *print(a is not c)*	*True* *False*

	return a True value.		

Membership Operators

A membership **operator** is used to test for a variable's value in a given string, tuple, or list.

For example:

a = 25

b = 35

c = [20, 25, 30, 45, 40]

if (a in list):

print("25 in list")

Result will print:

25 in list

To follow the examples in this section, create a new Python file under **Ch7Projects**, call it **ExMEMO.**

Clear any code in the **ExMEMO** file and type each of the following examples listed in the table below in Line 1.

Python has **two** basic **membership operators**.

Membership Operator	**Description**	**Example**	**Print Result**
in	The membership "in" operator will return True if the specific variable value is found in the list, tuple, or string. If the specified variable value is not found in the list, tuple, or string, it will be False.	*a = 8* *b = 10* *c = [2, 4, 6, 8, 10]* *if (a in c):* *print("8 in list")*	*8 in list*
not in	The membership "not in" operator will return True if the specific variable value is not found in the list, tuple, or string. If the specified variable value is found in the list, tuple, or string, it will be False.	*a = 8* *b = 10* *c = [2, 4, 6, 10]* *if (a not in c):* *print("8 not in list")*	*8 not in list*

Chapter 8: Functions, Keywords, and Identifiers

This chapter will explore some of the primary **functions** in Python and introduce keywords.

To follow the examples in this section, create a new Python project called **Ch8Projects**.

Functions

A **function** is a section or group of related code that is written to perform the desired task. By grouping code into specific functions, it helps to break a program into manageable chunks.

Functions are also handy for creating reusable code, which helps a programmer cut down on repetition and, in so doing, saves time and cuts down on errors.

To follow the examples in this section, create a new Python file under **Ch8Projects**, call it **ExFunctions.**

Function Syntax

keyword function_name(parameters):

"""docstring"""

statements

function_name(arguments)

Breakdown of the function syntax

Line #	Syntax	Description
1	*keyword*	The keyword that defines structure/syntax of the function. Keywords will be discussed in the next section of this chapter.
1	*function_name*	The name of the function. This can be a user-defined name or a built-in function.
1	*(parameters)*	A parameter is a variable being assigned to the function. There can be more than one variable/value defined at one time. For example: *(fname, lname)* A comma must separate each variable/value.
1	*:*	Statement terminator. A colon (:) denotes the end of the statement identifier.
2	*""""Docstring""""*	This line is an optional one. When writing large chunks of code, the programmer **comments** on what the code is for, how it works, etc.
3	*statements*	The body of the function is where the python statements to be executed

		when the function is called are defined.
4	*function_name*	This is the return statement that calls the function. It does not have to follow the function directly.
4	*(arguments)*	The value to be sent to the parameters set in the function when the function is called. If there are two parameters set in the function, two arguments must be sent when the function is called. For example: *(Mary, Green)* A comma must separate each variable/value.

Arguments and Parameters

When a function is defined, a **parameter(s)** can be set up so that information can be passed into a function.

When a function is called, an **argument(s)** are the values sent to the parameter(s) defined by a function.

The terms arguments and parameters are sometimes used interchangeably. This can get quite confusing, especially as a newbie programmer and because both instances are used to pass information into a function.

For example, for a list of first names, the function would look similar to the following:

def names(fname):

```
    print("First Names: " + fname)
first_name('Mary')
```

Result will print:

First Names: Mary

In the above code, it is assumed that *fname* is the set parameter for which *Mary* would be the value assigned to the *fname* parameter.

There can be more than one set parameter defined per function. Each parameter defined must be separated by a comma inside of the parentheses.

For example:

```
def names(fname, lname):
    print(fname + lname)
names('Mary ', 'Jones')
```

Result will print:

Mary Jones

When the function is called, two parameters must be sent to the function if it expects two parameters. For example, if the call function above only contained the argument "*Mary,*" the program would have ended with an error.

The function can be called more than once with different arguments passed to it.

For example:

```
def names(fname, lname):
    print(fname + lname)
names('Mary ', 'Jones')
names('Brian ', 'Monk')
```

names('Gary ', 'Brown')

Result will print:

Mary Jones

Brian Monk

Gary Brown

Things to Remember

A quick side note of things to remember, the first being that Python is a case sensitive language. This means that any variable, statement, keyword, or function used in Python has to be done in the case Python expects it to be in.

For example, when a variable name is set:

vname =

VName =

Whenever one of the above variables is referenced, they have to be done precisely as they are written.

For example:

vname = “Hi”

print(Vname)

Result will print in an error:

Traceback (most recent call last):

File "C:\Users\User\PycharmProjects\Ch3Projects\ExDataTypes.py", line 2, in <module>

print(Vname)

NameError: name 'Vname' is not defined

If the correct case for the variable is used in the *print()* function:

vname = "Hi"

print(vname)

Result will print:

Hi

Setting Default Parameter Values for a Function

Default parameter values can be set for function this is handy for when a call to the function does not pass through an argument value.

For example:

```
def funct1(hlang = "English"):
  print("My home language is " + hlang)
funct1()
```

Result will print:

My home language is English

When default parameters are set, they can be overridden by the call argument values.

For example:

```
def funct1(hlang = "English"):
  print("My home language is " + hlang)
funct1(German)
```

Result will print:

My home language is German

Using a List as an Argument for a Function

A function will accept all the known data types in Python. This means that a function will accept lists, dictionaries, numbers, strings, tuples, etc.

When a data type is passed to a function, the data type rules still apply. For instance, a string value must be enclosed in quotation marks; a number value does not have quotation marks, etc.

For example:

```
def funct1(animals):
  for z in animals:
    print(z)
animals = ['elephant', 'tiger', 'bear']
funct1(animals)
```

Result will print:

```
elephant
tiger
bear
```

Variable Scope

A scope of a variable is the section of the program where the variable is recognized.

A variable’s lifetime is the length of time the variable is held in the system's memory.

The examples below demonstrate the temporary nature of variables.

For example, variable a is set to a value of 10 on line 1:

```
a = 10
print(a)
```

Result will print:

```
10
```

That value of the variable a will only be set for the time the program is running or until variable a is assigned a new value.

For example:

a = 10

a = 20

print(a)

Result will print:

20

When a variable is defined inside of a function, it is not recognized outside of the function. Thus a variable defined within a function has a local scope and any similar variable set outside of the function's scope will not have any effect on it.

For example:

ef funcscope():

animals = "birds"

print("Animals defined within a function: " + animals)

animals = "dogs"

print("Animals defined outside of a function: " + animals)

funcscope()

Result will print:

Animals defined outside of a function: dogs

Animals defined within a function: birds

A variable is only valid while the program is executing, or until it is reassigned, a similar rule is applied to variables that have been defined within a function. The variable is only set and kept in memory while the function executes. Once the function ends, it moves on to another function or loops back to the current one. Any variables will be reset.

Python Built-In functions

There are two types of functions in Python, and they are:

- Python built-in functions

  These are Python functions that come with the Python package.

- User-defined functions

  These are functions created and defined by the programmer.

Python has many built-in functions that can be extended by adding on various modules or defining your own.

Python Built-In Functions

Built-In Function	Details	Example	Result
abs()	Absolute value	*a = -20* *print(abs(a))*	*20*
all()	If all values in an iterable list are true, it will return True. If one or more values in the iterable list are false, it will return	*list1 = [1, 3]* *a = all(list1)* *print(a)* *list1 = [0, 3]* *a = all(list1)* *print(a)*	*True* *False*

	False. If the object is empty the values will return as True.		
any()	If any values in an iterable list are true, it will return True. If one or more values in the iterable list are false, it will return False. If the object is empty the values will return as False.	*list1 = [1, 3]* *a = any(list1)* *print(a)* *list1 = [0, 1, 3]* *a = any(list1)* *print(a)*	*True* *True*
ascii()	Returns a readable string of characters. If the function comes across any non-ascii character in the string it will replace those characters with an escape	*a = ascii("Paint°\ Ball")* *print(a)*	*'Paint\xbo\\ Ball'*

	symbol.		
bin()	Converts an integer into binary code.	*a = bin(55)* *print(a)*	*0b110111*
bool()	Outputs the boolean value of an integer.	*a = bool("20")* *print(a)*	*True*
bytearray()	Outputs an array of bytes.	*a = bytearray(2)* *print(a)*	*bytearray(b'\x00\x00')*
bytes()	Outputs an array of bytes. The bytes() function converts values into bytes.	*a = bytes(2)* *print(a)*	*bytearray(b'\x00\x00')*
callable()	This function checks if an object can be called. If an object is callable, the function will return True. If an object is not callable the function will return False.	*def a():* *b = 25* *print(callable(a))*	*True*
chr()	Translate a	*a = chr(74)*	*J*

	unicode into its representing character.	*print(a)*	
complex()	Outputs complex real and imaginary parts of a selected number. The function can also convert a string into a complex number	*a = complex(4, -8)* *print(a)*	*(4-8j)*
dict()	Outputs a dictionary array.	*a = dict(animal = "tiger", species = "cat", genus = "panthera")* *print(a)*	*{'animal': 'tiger', 'species': 'cat', 'genus': 'panthera'}*
dir()	Prints out the methods as well as the properties of a specific function/object.	*class animal:* *type = "Cat"* *genus = "Panthera"* *print(dir(animal))*	['__class__', '__delattr__', '__dict__', '__dir__', '__doc__', '__eq__', '__format__', '__ge__', '__getattribute__', '__gt__', '__hash__', '__init__', '__init_subclass__', '__le__', '__lt__', '__module__',

			'__ne__', '__new__', '__reduce__', '__reduce_ex__', '__repr__', '__setattr__', '__sizeof__', '__str__', '__subclasshook__', '__weakref__', 'genus', 'type']
divmod()	The dividend value is divided by the divisor value to return the quotient and remainder of the division in tuple format.	a = divmod(7, 3) print(a)	(2, 1)
enumerate()	converts a collection of items in a tuple to enumerate values.	a = ('Tiger', 'Lion', 'Jaguar') b = enumerate(a) print(list(b))	[(0, 'Tiger'), (1, 'Lion'), (2, 'Jaguar')]
eval()	This function will evaluate an expression. If the expression	a = 'print(22)' eval(a)	22

	is a valid Python one the statement will run.		
exec()	This function will run/execute the referenced variable or blocks of code.	*a = 'animal = "panther"\nprint(animal)'* *exec(a)*	*panther*

filter()	This function is used to filter an iterable array of items.	*seasons = [1, 2, 3, 4]*
		def func1(a):
		  if a > 1:
		    return False
		  else:
		    return True
		winter = filter(func1, seasons)
		for a in winter:
		  print(a)
		filter(func1, winter)
		for a in season:
		  print(a)
			1

float()	Outputs a floating point number and will convert a number into a decimal.	*a = 4* *print(float(a))*	*4.0*
help()	This is a built in	*help(float)*	*class float(object)* *| float(x=0, /)*

	function that will activate the Python help utility.		| | *Convert a string or number to a floating point number, if possible.* | | *Methods defined here:* | | *__abs__(self, /)* | *abs(self)* | | *__add__(self, value, /)* | *Return self+value.*
hex()	This function converts a number into the hexadecimal format.	*num1 = 235* *print('hex =', hex(num1))*	*hex = 0xeb*
input()	This function is used to capture inputted data, converts it into a string and returns the value in the desired format.	*name = input("name: ")* *print("My name is ", name)*	*name: Mary* *My name is Mary*
int()	This function	*print(int(435.66))* *print(int("345"))*	*435* *345*

	returns an integer value of an object. This function will convert a string to a number.		
len()	This function will return the length of a specific object.	*animals = ["pig", "cow", "sheep"] print(len(animals))*	*3*
list()	The list function creates an iterable list.	*animals = ["pig", "cow", "sheep"] print(list(animals))*	*['pig', 'cow', 'sheep']*
max()	Returns the maximum value of a list.	*count1 = [5, 3, 8, 7] print(max(count1))*	*8*
min()	Returns the maximum value of a list.	*count1 = [5, 3, 8, 7] print(min(count1))*	*3*
next()	Prints out the next iterated value.	*farm = iter(["sheep", "cow", "horse"]) a = next(farm, "pig") print(a) a = next(farm, "pig") print(a) a = next(farm,*	*sheep cow horse pig*

		"pig") *print(a)* *a = next(farm, "pig")* *print(a)*	
object()	This function is used to return an empty object. Properties or methods cannot be added to this function.	*obj1 = object()* *print(type(obj1))* *print(dir(obj1))*	*<class 'object'>* ['__class__', '__delattr__', '__dir__', '__doc__', '__eq__', '__format__', '__ge__', '__getattribute__',
open()	This function is used to open and read a file Create a .txt file and call it readme.txt Type: Hello, I'm a text file. Save it as readme.txt **Parameters:** "r" — read "a" — append "w" — write "x" — exclusive write	*f = open("readme.txt", "r")* *print(f.read())*	*Hello, I'm a text file*

	"t" – text mode "b" – binary mode "+" – read & write		
pow()	This function returns the power of a number to a specified value.	*print(pow(9, 2))*	*81*
print()	This function prints out the contents of the brackets.	*print("Hello, this is me")*	*Hello, this is me*
range()	This function will return an immutable sequence of numbers starting from 0. A start and stop value determines the range of numbers in the sequence, which is incremente	*range1 = range(6, 10)* *for a in range1:* *print(a)*	*6* *7* *8* *9*

	d by 1.		
reversed( )	This function returns list values in reverse order.	*test1 = ["a", "b", "c", "d"]* *test2 = reversed(test1)* *for a in test2:* *print(a)*	*d* *c* *b* *a*
round()	Rounds a number to the nearest whole number.	*g = 234.9876565* *print(round(g))*	*235*
set()	This function will create an unordered set of items/value s.	*g = set(("pigs", "cows", "hens", "sheep"))* *print(g)*	*{'sheep', 'pigs', 'cows', 'hens'}*
slice()	This function will slice a list, tuple, dictionary, or string.	*g = ("pigs", "cows", "hens", "sheep")* *h = slice(1)* *print(g[h])*	*('pigs',)*
sorted()	This function storts a list of iterable values It can be used for both strings and numbers. Strings will	*g = ("pigs", "cows", "hens", "sheep")* *h = sorted(g)* *print(g)* *i = (8, 2, 6, 4)* *j = sorted(i)* *print(j)*	*('pigs', 'cows', 'hens', 'sheep')* *[2, 4, 6, 8]*

	be sorted alphabetically and numbers, numerically.		
str()	This function will convert a number value into a string value.	*g = (2563)* *print(str(g))*	*"2563"*
sum()	This function will return the sum of an iterable table.	*g = (5, 10, 15)* *print(sum(g))*	*30*
tuple()	This function creates a tuple array of values.	*g = tuple(("hen", "cow", "horse"))* *print(g)*	*('hen', 'cow', 'horse')*
zip()	This function creates what is called a zip object. When there are two tuples of information, the zip function will pair the first tuple	*animal = ("Lion", "Whale", "Eagle")* *atype = ("land", "sea", "sky")* *c = zip(animal, atype)* *print(tuple(c))*	*(('Lion', 'land'), ('Whale', 'sea'), ('Eagle', 'sky'))*

	with the second one.		

Keywords

A **keyword** is a reserved word in Python used for specific purposes, such as defining a program's structure and syntax.

Python has thirty-three or more (depending on the version of Python) keywords. It is good to know these keywords as they cannot be used as variable names or user-defined function names.

There are only a few keywords used in this course; as you advance to an intermediate level, you will learn more about the keywords listed below. For the beginner course, you need to be aware of them. You do not accidentally use a keyword to define a variable or function.

There are only three Python keywords that start with a capital letter (True, Fales, None); the rest begin with a lowercase letter.

To follow the examples in this section, create a new Python file under **Ch8Projects**, call it **ExKeywords**.

Python Keywords

Keyword	Description
and	Logical "and" operator
as	To create an alias when importing or

	executing modules.
assert	Used for debugging a program.
async	This keyword can be used to write concurrent code.
await	This keyword can be used to write concurrent code.
break	This keyword is used to break a while or for loop.
class	This keyword is used to define a new user-defined Python class.
continue	This keyword is used to break current loop processes and not the entire loop process.
def	This keyword is used to define a function (user-defined function).
del	This keyword is used to del a value/object.
elif	This keyword is used for conditional operations and is short for 'else if'.
else	This keyword is used for conditional operations (if, then, else).
except	This is an exception keyword and is used

	for catching exceptions in the Python code
False	This keyword is a Python truth value that results from logical or comparison operations.
finally	This keyword is used in debugging alongside the try - except keywords.
for	This keyword is used in Python loops.
from	This keyword is used with the import keyword to import specific aspects of a module into Python.
global	This keyword defines a variable that is contained within a function to be global. This means the variable can be seen outside of the function scope.
if	This keyword is used for conditional operations (if, then, else).
import	These keyword imports modules into Python.
in	This keyword tests whether a value exists inside a string, tuple, list, or dictionary.
is	This keyword is to test whether or not an object is referenced by more than one variable.

lambda	This keyword is used to create a function with no name, otherwise referred to as an anonymous function.
None	This keyword represents a null value or the absence of a value.
nonlocal	This keyword defines a variable that is contained within a nested function to be non-local. This means the variable can be seen outside of the function scope.
not	Logical "not" operator
or	Logical "or" operator
pass	This keyword is used to pass over a function instead of executing it.
raise	This is an exception keyword and is used for catching exceptions in the Python code.
return	This keyword is used to return to a value by exiting the executed function.
True	This keyword is a Python truth value that results from logical or comparison operations.
try	This is an exception keyword and is used for catching exceptions in the Python

	code.
while	This keyword is used to loop a statement until a specific condition is met, and the loop ends.
with	This keyword is used in conjunction with the context manager to wrap code execution within a particular method.
yield	This statement is used to generate one value to store in memory at a time.

Identifier

Identifiers are the names given to various elements in a Python program, such as variables, lists, tuples, dictionaries, class, functions, and strings. Throughout this tutorial, you have been assigned identifiers each time you created a variable.

For example:

animals = "Horse", "Donkey", "Zebra"

The above code "**animals**" is the string's identifier that follows the "=" sign.

Rules for Creating Identifiers

There are a few rules to follow when creating identifiers to ensure that your code meets the Python standards and runs correctly.

- There is no restriction on the length of an identifier, but it is better to keep them as short as possible.

- It is good practice to give an identifier a name that best represents what it is being used for; for instance, a list of animals could be called animal_list.
- Python is case sensitive, as discussed throughout this tutorial, as it is a critical point to remember.
- Use a standard case to set identifiers throughout your program.
- Use the underscore (_) to separate identifier names (animal_names) and not a space.
- Do not start an identifier with a digit (1animal) as this is invalid and will throughout an error.
- An identifier can contain letters and numbers (animal1, a1nimal, an1mal).
- A keyword cannot be an identifier.
- Special characters/symbols cannot be used as or part of an identifier.
    - ☐ Special symbols/characters include:

      \ / @ # $ % ^ & * ! ~ ? > < , . = + - : " ' ; [ ] ( ) { }

Chapter 9: Loops

This chapter will explore **loops** in Python and how they work.

To follow the examples in this section, create a new Python project called **Ch10Projects**.

What is a Loop?

In programming, a loop is a block of code that will repeat over and over until a certain criteria/condition is met.

Some of the main functions of using loops in a program are to repeat functions when necessary, add values to variables, and cycle through various programming steps.

All of the modern programming languages, available in the marketplace today support loops, including Python.

Python Loops

Python programming supports two types of loops, the **for** loop and the **while** loop.

In both the for and while loops, Python uses indentation to group blocks of code as it does for statements and functions.

For Loop

To follow the examples in this section, create a new Python file under **Ch9Projects**, call it **ExForLoop**.

The for loop is used to iterate through a string, tuple, dictionary, set, or list.

Clear any code in the **ExForLoop** file and type the following in Line 1:

```
animals = ['horse', 'cow', 'sheep', 'goat']
for b in animals:
  print(b)
```

Result will print:

horse

cow

sheep

goat

Statements can be run on each item in a sequence, and no indexing variable needs to be set for a for loop.

Clear any code in the ExForLoop file and type the following in Line 1:

```
for b in "sheep":
  print(b)
```

Result will print:

Result will print:

Result will print:

s

h

e

e

p

While Loop

To follow the examples in this section, create a new Python file under **Ch9Projects**, call it **ExWhileLoop**.

The **while loop** is used to loop blocks of a statement. As long as the set condition for the while loop remains **True,** the loop will execute.

As soon as the set condition becomes **False**, the while loop will exit to the next block of code. You have to be careful with the while loop and ensure a set condition to be met otherwise; you could send the program into an infinite loop.

For example, set a counter to print a particular code block for a given amount of time in the following exercise.

Clear any code in the **ExWhileLoop** file and type the following in Line 1:

Write the following code:

```
print_counter = 0
while(print_counter < 5):
    print("While loop number: ", print_counter)
```

Result will print the following over and over in an infinite loop:

```
While loop number: 0
While loop number: 0
```

While you are testing the script in PyCharm, it is easy to stop the script. But for a program being run in a live environment, you could potentially bring the entire system to a standstill.

If you write the same code, with both a counter and increment that counter for each loop as per the code below, you will get a better result.

```
print_counter = 0
```

```
while(print_counter < 5):
   print_counter = print_counter + 1
   print("While loop number: ", print_counter)
```

Result will print:

While loop number: 1

While loop number: 2

While loop number: 3

While loop number: 4

While loop number: 5

Loop Control Statements

A **loop control statement** is what is used to change the execution of a program loop. A control statement either instructs the program to return to the beginning of the loop or exit the loop and end sequence.

There is three main loop control statement in Python:

- Break
- Continue
- Pass

Break Statement

When exiting a **for** or **while loop,** the **break statement** is used in Python. When specific set criteria are reached, the break statement will exit the loop and continue onto the next code section. In other words, it will not loop through the entire contents of a statement.

For example, if you had an extensive list of numbers, you could print all the odd numbers, and instead of printing out all the even ones before exiting the loop, you can set the criteria to only print odd numbers.

To follow the examples in this section, create a new Python file under **Ch9Projects**, call it **ExBreakControl**.

Clear any code in the **ExBreakControl** file and type the following in Line 1:

```
turns = 0
while True:
    print(turns)
    turns += 1
    if turns == 5:
        break       # When the loop has run through 5 times the loop will exit
print("\n")  # Next piece of code executed directly after the break statement
for b in range(1, 10):
    if b % 2 == 0:
        continue
    print(b)
```

Result will print:

```
0
1
2
3
4
1
3
5
7
9
```

Continue Statement

To **return to the beginning** of a **for** or **while loop,** the **continue statement** is used in Python.

There may be one or two items that do not need to be processed in a loop. You do not want to break the loop, and you want to skip them but also continue processing the items in the loop.

For example, you may have a list of colors, but you want to exclude green and orange. You would write in an exception to exclude those items but continue the loop.

To follow the examples in this section, create a new Python file under **Ch9Projects**, call it **ExContinueControl**.

Clear any code in the **ExContinueControl** file and type the following in Line 1:

```
for b in range(1, 5):
    if b == 2:
        continue
    print(b)
```

Result will print:

```
1
3
4
```

You will notice in the above example that the 2 was left out of the range.

When setting a range, it is essential to note that the range's scope will include the first number in the range but exclude the range's last number.

In the example above, the range was set from 1 to 5. The computer will see it as 1, 2, 3, 4.

Pass Statement

To **pass over** a **for** or **while loop,** the **pass statement** is used in Python. The loop should never be empty, and when the program comes across a loop, it will execute it.

If you have a blank loop or loop that you want the code to later, you must include a pass statement.

To follow the examples in this section, create a new Python file under **Ch9Projects**, call it **ExPassControl**.

Clear any code in the **ExPassControl** file and type the following in Line 1:

```
while x = True:
 pass
animals = ["dog", "cat", "monkey"]
for g in animals:
pass
```

Using If, Else, and Elif With Loops

Conditional keywords, **if**, **else**, and **elif** can be used with for and while loops.

To follow the examples in this section, create a new Python file under **Ch9Projects**, call it **ExIfElseElif**.

If

The **if** keyword is used to test for logical conditions between two items.

For example:

Clear any code in the **ExIfElseElif** file and type the following in Line 1:

```
counter = 0
while counter < 10:
  counter += 1
  if counter == 8:
    continue
  print(counter)
```

Result will print:

```
1
2
3
4
5
6
7
9
10
```

Clear any code in the **ExIfElseElif** file and type the following in Line 1:

```
for a in range(10):
  if a == 8:
    break
  print(a)
```

Result will print:

```
1
2
3
4
5
6
7
9
100
1
2
3
4
5
6
7
```

Else

The **else** keyword is used once the condition of the loop is no longer true

For example:

Clear any code in the **ExIfElseElif** file and type the following in Line 1:

```
counter = 0
while counter < 5:
```

```
    print(counter)
    counter += 1
else:
    print("counter exceeds the maximum number of counts")
```

Result will print:

```
1
2
3
4
counter exceeds the maximum number of counts
```

Clear any code in the **ExIfElseElif** file and type the following in Line 1:

```
or b in range(10):
    if b == 7:
        break
    print(b)
else:
    print("range exceeded")
```

Result will print:

```
0
1
2
3
4
5
6
```

Elif

The **elif** keyword is the same as **else if** and is used when an expression must include multiple if ... else conditions.

The following code forms part of the HangMan project code from a later Chapter of this book where you will get to put your new Python coding knowledge into practice. The following code gives you an example of how the **elif** keyword can be used with a **while** and **for** loop.

```
while not picked and chances > 0:
    guess = input("Input either a word or a letter to start: ").upper()
    if len(guess) == 1 and guess.isalpha():
        if guess in picked_letters:
            print("You already used that character", guess)
        elif guess not in pick:
            print(guess, "is incorrect")
            chances -= 1
            picked_letters.append(guess)
        else:
            print("Well done,", guess, "is correct")
            picked_letters.append(guess)
            index_words = list(comp_word)
            indexing = [i for i, letter in enumerate(pick) if letter == guess]
```

Nested Loops

Nested loops are loops within a loop.

To follow the examples in this section, create a new Python file under **Ch9Projects**, call it **ExNestedLoops**.

Clear any code in the **ExLoop** file and type the following in Line 1:

```
for a in range(1, 8):
  for b in range(a):
    print(1, end = " ")
  print()
```

Result will print:

```
1 1 1
1 1 1 1
1 1 1 1 1
1 1 1 1 1 1
1 1 1 1 1 1 1
```

Clear any code in the **ExLoop** file and type the following in Line 1:

```
c = 2
while(c < 50):
  a = 2
  while(a <= (c / a)):
    if not(c % a): break
    a += 1
  if (a == c / a):
```

```
        print(a, "must be a prime number")
    c += 1
print("The end")
```

Result will print:

```
2 must be a prime number
3 must be a prime number
5 must be a prime number
7 must be a prime number
The end
```

Chapter 10: Python Package Managers and Modules

While working with Python, you will hear the word Python packages, modules, and pip being mentioned. This chapter will give a brief introduction and overview of what they are and why they are essential to Python.

Python Modules

Modules are Python code library files that contain various Python definitions and statements.

For example, in some of the projects you will be doing at the end of the book, you will be working with the "**random**" module. When this module is imported into Python, it increases Python's functionality.

The "random" module, for example, extends the programming scope to allow the program to perform rand operations such as:

- **randint()** – The program randomly selects an integer value between a set range of numbers.
- **random()** – The program randomly selects a float value between 0 and 1.
- **choice()** – The program randomly chooses an item from a given selection of items.

There are many more methods included in the "random" module. Python also has over two-hundred different modules to choose from.

Python modules are saved as .py files (standard Python files). Python allows for user-created modules. In the Hangman project, you will be creating a simple wordlist module. Creating modules is something you will learn in a more advanced Python course.

To follow the examples in this section, create a new Python file under **Ch10Projects**, call it **ExModules**.

The for loop is used to iterate through a string, tuple, dictionary, set, or list.

To import modules into Python code, the syntax is:

import <module>

For example:

import random

To retrieve one of the methods from the imported module you the syntax is:

from <module> import <method>

For example:

from random import randint

To learn about the imported package/module, type the following code:

import random

help(random)

print(help)

The help file will display the following (note the following print out is only a snippet of the full help file).

Result will print:

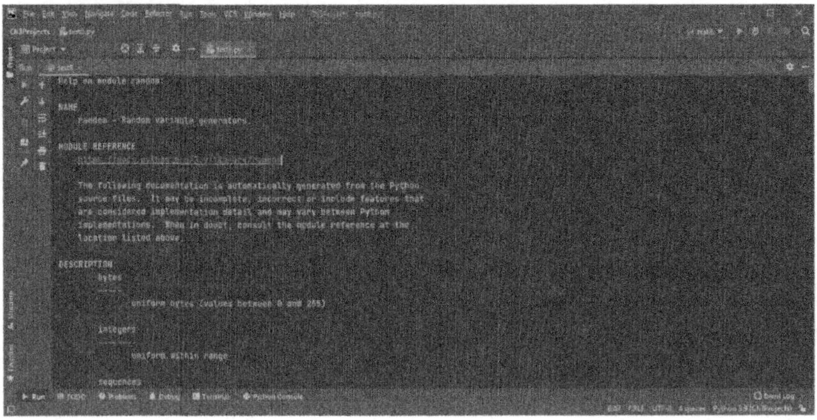

Python Packages

A Python package is a collection of related Python modules. In other words, it is a directory of Python modules that also contains some additional information, such as the _init_.py file.

Packages can become quite complicated as they can be nested within each other to form multi-depth libraries of Python modules.

Python Package Managers

Python packages can become quite cumbersome to manage and download as they can have their own dependencies. If you are unaware of these, a needed package may not get installed correctly, which would affect your program functionality.

Package managers help to manage these dependencies. Python is popular because of its extensive library of packages to work with.

Popular Python Package Managers

Package Manager	Description
Anaconda	Anaconda comes pre-installed with Python and many other development tools that make it popular for data science, engineering, analytics, big data, and more. Anaconda works with Pip, which makes the package capable of importing all the Python libraries available. Anaconda separates all its projects in a virtual environment to isolate different projects and versions of projects.
Pip	Pip is Python's Package manager. Pip comes with Python 3.4 and above.

Working with Pip

Pip works best with your operating system command line.

Open the "**Command Prompt**" by either typing **cmd** into the search bar or finding the Command Prompt in the "**Programs Menu**" of Windows.

Once the Command Prompt screen is open, navigate to the following directory:

cd\Users\UserName\AppData\Local\Programs\Python\Python39\Scripts

The first thing to do is check to see if Pip has been installed. Type the following in a clean line of the Command Prompt screen.

pip --version

Result will print:

pip 20.2.3 from
c:\users\username\appdata\local\programs\python\python39\lib\site-packages\pip (python 3.9)

The version is pip 20.2.3 / pip (python 3.9)

The latest versions of Python should automatically install Pip during the Python application installation.

To download a package with Pip, type the following in the Command Prompt window:

pip install -U scikit-learn

As Scikit-Learn starts to unpack and install you will notice all the files, modules, and packages associated with the package. This demonstrates how important a package manager is to ensure that all the files associated with a package and its modules not only get downloaded correctly but installed correctly as well.

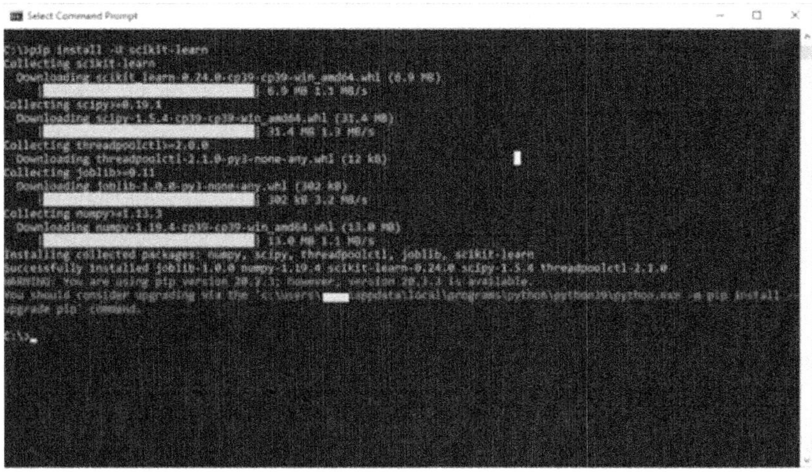

Once the package has been successfully collected, downloaded, and installed, the Pip package manager will return a "Successfully installed" message.

In the image above, you will notice that once the package was installed, Pip returned a warning. This was to let the programmer know that there is a newer version of Pip available.

Pip can be upgraded by typing the following in the Command Prompt window:

python -m pip install --upgrade pip

Packages can be uninstalled using Pip by typing the following in the Command Prompt window:

pip uninstall scikit-learn

There will be a warning message asking the programmer if they are sure they want to uninstall the package and the associated files that will be uninstalled with the program.

At the “Proceed (y/n)?” prompt, type y, and the package will be uninstalled.

To check what packages are currently installed on the system, type the following at the Command Prompt screen:

pip list

A screen similar to the one below, will be displayed listing all the installed packages.

You can find more Python packages at the PyPI website using the following URL:

https://pypi.org/

Chapter 11: Python Practice Projects

Now that you know Python's basic structure and code, it is time to put what you have learned into practice with these two Python for beginners' projects.

Create a new Python project called **Projects.**

Python Project 1 – Mad Libs Generator

A fun, straightforward program to get started with is a Mad Libs Generator game.

For those who have never played the Mad Libs game, here is a quick overview:

Mad Libs is a game that consists of word cards and sentence cards.

The word cards come with words on them that consist of a mix of:

- Verb
- Noun
- Plural noun
- Adverbs
- Adjectives

Each player gets seven word cards to start, and they must pick a word from their card to fill in the blanks of the chosen sentence card.

The game turns into a fun silly worded sentence game that happens to be a great project to create as your first Python coding project.

Have fun with the code, switch it out a bit and see if you can add to it and make it a little more complicated the more advanced you get.

Create a new Python file in the Python **Projects** folder and call it **MadLibs**.

Clear any code in the **MadLibs** file and type the following code from Line 1:

```
# Simple Mad Libs Generator Game
# prompt the user to enter a noun
noun1 = input("Please enter a noun: ")
# prompt the user to enter an adjective
adjective = input("Please enter an adjective: ")
# prompt the person for another noun
noun2 = input("Please enter another noun: ")
# Print Mad Libs Sentence
print(noun1 + " are faster than the wind")
print("Ducks are " + adjective)
print("Don't run with a " + noun2)
```

Result will print:

```
Please enter a noun: horses
Please enter an adjective: pink
Please enter another noun: fork
horses are faster than the wind
Ducks are pink
Don't run with a fork
```

Python Project 2 — Calculator

In this project, you are going to create a simple, functional calculator. This project will help you become familiar with the if - else - elif functions.

As with all the tutorial projects, try and improve on the code, customize it, and make yourself an even fancier calculator. The more you practice and have a hands-on approach, the faster you will learn. Remember, it is good to make mistakes in programming; that is how we learn.

Create a new Python file in the Python **Projects** folder and call it **Calculator**.

Clear any code in the **Calculator** file and type the following code from Line 1:

```
# A simple Python calculator

# define a function to add two numbers

def add(a, b):

   return a + b

# define a function to divide two numbers

def divide(a, b):

   return a / b

# define a function to multiply two numbers

def multiply(a, b):

   return a * b

# define a function to subtract two numbers

def subtract(a, b):

   return a - b

print("Select Calculator Operation:")
```

```
print("1.Add")
print("2.Divide")
print("3.Multiply")
print("4.Subtract")
while True:
    # input arithmetic operator choice
    select1 = input("Enter 1, 2, 3, or 4: ")
    # Validate the choice selection
    if select1 in ('1', '2', '3', '4'):
        num1 = float(input("First number: "))
        num2 = float(input("Second number: "))
        if select1 == '1':
            print(num1, "+", num2, "=", add(num1, num2))
        elif select1 == '2':
            print(num1, "/", num2, "=", divide(num1, num2))
        elif select1 == '3':
            print(num1, "*", num2, "=", multiply(num1,
num2))
        elif select1 == '4':
            print(num1, "-", num2, "=", divide(num1, num2))
        break
    else:
        print("Please make a valid choice")
```

Result will print:

Select Calculator Operation:

1.Add

2.Divide

3.Multiply

4.Subtract

Enter 1, 2, 3, or 4: 4

First number: 2

Second number: 3

2.0 - 3.0 = 0.6666666666666666

Python Project 3 — Guess the Number

This is a simple number guessing game where player1 will select a range of numbers, for example, from 1 to 50.

The computer will select a random number from within the user-defined range; for the sake of this explanation, assume the computer selected 5.

Player1 has five chances to guess the number 5.

If player1 guesses a number greater than 5, the program will return:

"Sorry, the number is less than that."

If player1 guesses a number less than 5, the program will return:

"Sorry, the number is greater than that."

If player1 guesses the number, the program will return:

"Well done! You guessed the number in x amount of attempts."

If player1 does not get the answer right in 5 turns, the program will return:

"Sorry, but you are out of guesses; better luck next time."

Create a new Python file in the Python **Projects** folder and call it **NumberGame**.

Clear any code in the **Number** file and type the following code from Line 1:

 # Number guessing game

 # import the random and math modules to run the game

import random

 # import the randint() function from the random module

 # randint() function makes it possible for the computer

 # to generate random numbers

from random import randint

 # the math module allows for extra math function support

 # needed for this game

import math

 # player1 must input the lower value of the range

low = int(input("Enter the lower number of the ranger: "))

 # player1 must input the highest value of the range

```
high = int(input("Enter the highest number of the range: "))

# computer must generate a random number between the set range

rn = random.randint(low, high)

print("\n\tYou have ", round(math.log(high - low + 1, 2)),"guesses left!\n")

# start the counter for logging the number of guesses

count = 0

# number of guesses will depend on the number range

#

while count < math.log(high - low + 1, 2):

    count += 1

    # prompt player1 to guess the random number

    pick = int(input("The number is: "))


    # the if statement will start the conditional

    # testing to determine if player1 guessed correctly

    if rn == pick:

        print("Well done! You guessed the number in ", count, "attempts")

        # if player1 guesses the number on the first try, the loop will end

        break

    elif rn > pick:

        print("Sorry the number is less than that")
```

```
    elif rn < pick:
      print("Sorry the number is higher than that")


# if player1 has gone over the allowed number of attempts
# the program will execute the following block of code
if count >= math.log(high - low + 1, 2):
  print("\nThe computer chose the following number %d"%rn)
  print("\tSorry but you ran out attempts to guess the number")
```

Result will print:

Enter the lower number of the ranger: 1

Enter the highest number of the range: 10

You have 3 guesses left!

The number is: 2

Well done! You guessed the number in 1 attempts.

Python Project 4 – Hangman Game

Hangman is a word guessing game where the computer randomly picks a word from a pre-written list of words. There is a picture of a gallows, and for each wrong answer, the image will be incremented with a rope and head, then a rope - head- 1 am, then a rope - head - 2 arms, etc. until the word is either guessed or the player is hung.

As there are only so many body parts, the player is limited to the number of chances they get to guess the correct word.

The program will allow the player to input an entire word and guess the word character by character. For each incorrect word, the player will lose a turn/life; one more body part is added to the gallows until there are no tries left.

You will create two Python files in the **Projects** folder. One will be the Hangman code, and the other one will be a list of Hangman words for the computer to randomly select.

There is some intermediate coding in this program, but it is not hard to follow or figure out. This project and project 5 (the next project) are an excellent passageway into intermediate Python.

Create two new Python files in the Python **Projects** folder:

- Call the first file **hmwords**.
- Call the second file **hangman**.

Clear any code in the **hmwords and hangman** files.

hmwords.py File

Start with **hmwords.py** file and type the following code from Line 1:

```
hm_words= [
    'donkey',
    'animal',
    'stutter',
    'money',
    'roller',
    'tandem',
    'flower',
    'backpack',
    'flower',
```

```
        'mask',
        'honey',
        'train',
        'bike',
        'soup',
        'dollar',
        'pound'
    ]
```

Result will print:

Save the file and run the code. There will be no results.

This file can be left open while you work on the HangMan file.

You can add as many words to the **hmwords.py** file as you want to.

A list of HangMan words can be found on the website **Hangman words.** This website can be found at the following URL :

https://www.hangmanwords.com/

hangman.py File

In the **hangman.py** file, type the following code from Line 1 – Note: the text written in bold is for explanation only and should not be written into the program:

```
# Game of hangman
import random
from hmwords import hm_words
def pick_word():
    pick = random.choice(hm_words)
```

```
    return pick.upper()
def game(pick):
    comp_word = "_" * len(pick)
    picked = False #
    picked_letters = []
    picked_words = []
    chances = 6
    print("Start playing Hangman!")
    print(hangman_gallows(chances))
    print(comp_word)
    print("\n")
    while not picked and chances > 0:
        guess = input("Input either a word or a letter to
start: ").upper()
        if len(guess) == 1 and guess.isalpha():
            if guess in picked_letters:
                print("You already used that character",
guess)
            elif guess not in pick:
                print(guess, "is incorrect")
                chances -= 1
                picked_letters.append(guess)
            else:
                print("Well done,", guess, "is correct")
                picked_letters.append(guess)
```

```
            index_words = list(comp_word)
            indexing = [i for i, letter in enumerate(pick) if letter == guess]
            for index in indexing:
                index_words[index] = guess
            comp_word = "".join(index_words)
            if "_" not in comp_word:
                picked = True
    elif len(guess) == len(pick) and guess.isalpha():
        if guess in picked_words:
            print(guess, " has already been used")
        elif guess != pick:
            print(guess, "is not the word.")
            chances -= 1
            picked_words.append(guess)
        else:
            picked = True
            comp_word = pick
    else:
        print("Your selection is not valid!")
    print(hangman_gallows(chances))
    print(comp_word)
    print("\n")
if picked:
```

```
    print("Well done you have successfully guessed the word!")
  else:
    print("Oops looks like you ran out of tries. The word was " + pick)
def hangman_gallows(chances):
  gallows = [ # full hangman: head, body, arms, and legs
    """
      --------
      |   |
      |   O
      |  \\|/
      |   |
      |  / \\
      -
    """,
    # head, body, arms, and one leg
    """
      --------
      |   |
      |   O
      |  \\|/
      |   |
      |  /
      -
```

```
""",
# head, body, and arms
"""
  --------
  |     |
  |     O
  |    \|/
  |     |
  |
  -
""",
# head, body, and one arm
"""
  --------
  |     |
  |     O
  |    \|
  |     |
  |
  -
""",
# head and body
"""
  --------
  |     |
```

```
  |    O
  |    |
  |    |
  |
  -
""",
# head
"""
  --------
  |    |
  |    O
  |
  |
  |
  -
""",
# initial state gallows
"""
  --------
  |    |
  |
  |
  |
  |
  -
```

```
    """
  ]
  return gallows[chances]
def main():
  pick = pick_word()
  game(pick)
  while input("Would you like to have another try? (Y/N) ").upper() == "Y":
    pick = pick_word()
    game(pick)
if __name__ == "__main__":
  main()
```

Result will print for selecting character by character:

```
Start playing Hangman!
   --------
   |   |
   |
   |
   |
   |
   -

________
Input either a word or a letter to start: d
Well done, D is correct
   --------
```

```
        |   |
        |
        |
        |
        |
        -
____D__
Input either a word or a letter to start: a
Well done, A is correct
        --------
        |   |
        |
        |
        |
        |
        -

_A_D__
Input either a word or a letter to start: r
R is incorrect
        --------
        |   |
        |   O
        |
        |
```

```
    |
    -
_A_D__
Input either a word or a letter to start: g
G is incorrect
    --------
    |   |
    |   O
    |   |
    |   |
    |
    -
_A_D__
Input either a word or a letter to start: h
H is incorrect
    --------
    |   |
    |   O
    |  \|
    |   |
    |
    -
_A_D__
Input either a word or a letter to start: e
Well done, E is correct
```

```
--------
|    |
|    O
|   \|
|    |
|
-
```

_A_DE_

Input either a word or a letter to start: g

You already used that character G

```
--------
|    |
|    O
|   \|
|    |
|
-
```

_A_DE_

Input either a word or a letter to start: n

Well done, N is correct

```
--------
|    |
|    O
|   \|
|    |
```

```
        |

        -
```

ANDE

Input either a word or a letter to start: r

You already used that character R

```
        --------

        |    |

        |    O

        |   \|

        |    |

        |

        -
```

ANDE

Input either a word or a letter to start: t

Well done, T is correct

```
        --------

        |    |

        |    O

        |   \|

        |    |

        |

        -
```

TANDE_

Input either a word or a letter to start: m

Well done, M is correct

```
--------
|    |
|    O
|   \|
|    |
|
-
```

TANDEM

Well done, you have successfully guessed the word!

Would you like to have another try? (Y/N) y

Start playing Hangman!

```
--------
|    |
|
|
|
|
-
```

____

Input either a word or a letter to start: a

A is incorrect

```
--------
|    |
|    O
|
```

```
   |
   |
   -
_______
Input either a word or a letter to start: d
D is incorrect
   --------
   |    |
   |    O
   |    |
   |    |
   |
   -
_______
Input either a word or a letter to start: s
S is incorrect
   --------
   |    |
   |    O
   |   \|
   |    |
   |
   -
_______
Input either a word or a letter to start: f
```

F is incorrect

```
--------
|    |
|    O
|   \|/
|    |
|
-
```

______

Input either a word or a letter to start: Your selection is not valid!

```
--------
|    |
|    O
|   \|/
|    |
|
-
```

______

Input either a word or a letter to start: h

H is incorrect

```
--------
|    |
|    O
|   \|/
```

```
        |   |
        |  /
        -
______
```

Input either a word or a letter to start: b

B is incorrect

```
        --------
        |   |
        |   O
        |  \|/
        |   |
        |  / \
        -
______
```

Oops looks like you ran out of tries. The word was MONEY.

Would you like to have another try? (Y/N)

Result will print for selecting character by character:

Start playing Hangman!

```
        --------
        |   |
        |
        |
        |
        |
```

```
    -

________

Input either a word or a letter to start: flower

    --------

    |    |

    |

    |

    |

    |

    -

FLOWER

Well done, you have successfully guessed the word!

Would you like to have another try? (Y/N) y

Start playing Hangman!

    --------

    |    |

    |

    |

    |

    |

    -

_______

Input either a word or a letter to start: cat

Your selection is not valid!
```

The game will continue until the player runs out of moves and is hung, or guessed the correct word.

As you can see, with the basic instruction you have learned in this tutorial, you can create some fantastic code.

Python Project 5 — Rock - Paper - Scissors Game

In this project, you will import a few modules to create a game of rock - paper - scissors.

The project is an introduction to Python intermediate level as it has a few intermediate level features in it. It is an excellent project to phase into the next level of Python programming.

Create a new Python file in the Python **Projects** folder and call it **RPS**.

Clear any code in the **RPS** file and type the following code from Line 1:

```
# Rock - paper - scissors game

# import random module to enable the computer to play against the user

import random

from random import randint

# print out a help file to instruct the user on how to play

print("Rock - Paper - Scissors Rules: \n"

    + "Paper covers Rock. Paper wins. \n"

    + "Rock smashes Scissors. Rock wins \n"

    + "Scissors cuts Paper. Scissor wins \n")
```

```
# selection menu for game play
while True:
    print("Choose your game play: \n 1. Rock \n 2. Paper \n 3. Scissor \n")
    # user input - player1
    player1 = int(input("Player 1 it's your turn: "))
    # Create a loop to ensure the player's selection is a valid entry
    while player1 > 3 or player1 < 1:
        player1 = int(input("enter valid input: "))
    # define the player1_choice variable
    if player1 == 1:
        player1_choice = 'Rock'
    elif player1 == 2:
        player1_choice = 'Paper'
    else:
        player1_choice = 'Scissor'
    # print out player1's game play option
    print("Player 1 you chose: " + player1_choice)
    print("\nPlease allow your computer to make its choice ...")
    # the random module contains randint method
    # the randint method will return random integer values
    # these random integer values are set between two specific values
```

```
# the values for this set will be 1, 2, and 3 — (1, 3)
player2 = random.randint(1, 3)
# the computers choice will loop until it equals player1 choice
while player2 == player1:
  player2 = random.randint(1, 3)
# initialize value of comp_choice_name
# variable corresponding to the choice value
if player2 == 1:
  player2_choice = 'Rock'
elif player2 == 2:
  player2_choice = 'Paper'
else:
  player2_choice = 'Scissor'
print("The computer has chosen: " + player2_choice)
print(player1_choice + " vs " + player2_choice)
# game condition testing to find the winner
if ((player1 == 1 and player2 == 2) or
    (player1 == 2 and player2 == 1)):
  print("Paper wins this round => ", end="")
  result = "Paper"
elif ((player1 == 1 and player2 == 3) or
    (player1 == 3 and player2 == 1)):
  print("Rock wins this round =>", end="")
  result = "Rock"
```

```
  else:
    print("Scissor wins this round =>", end="")
    result = "scissor"
  # Printing either user or computer wins
  if result == player1_choice:
    print(" Player 1 you are the Winner")
  else:
    print("The Computer Wins! Better Luck next time!")
  print("Would you like to play again? (Y/N)")
  ans = input()
  # if player1 wants to print again, the game will start again (Y)
  # if player1 wants to stop the game will end (N)
  if ans == 'n' or ans == 'N':
    break
# if player1 has decided to quit playing
# the game will print thank you for playing
print("\nThanks for playing, I hope we can play again!")
```

Result will print:

Rock - Paper - Scissors Rules:

Paper covers Rock. Paper wins.

Rock smashes Scissors. Rock wins

Scissors cuts Paper. Scissor wins

Choose your game play:

1. Rock

2. Paper

3. Scissor

Player 1 it's your turn: 1

Player 1 you chose: Rock

Please allow your computer to make its choice ...

The computer has chosen: Paper

Rock vs. Paper

Paper wins this round => The Computer Wins! Better luck next time!

Would you like to play again? (Y/N)

Project 6— Tic Tac Toe

In this project you are going to create a game of tic tac toe.

As per the last two projects above, this project is an introduction to Python intermediate level as it has a few intermediate-level features in it.

Create a new Python file in the Python **Projects** folder and call it **TicTacToe**.

You are going to have to install a module named Numpy to create the dimensional array for the game.

In PyCharm go to the top menu and select the "File" menu option.

From the "File" menu drop-down list choose the "Settings" option.

Click on the "Project: TicTacToe" option.

On the screen on the right, click on the "Python Interpreter" option.

Click the "+" button on the bottom left of the "Package" box.

Type "numpy" in the search bar.

Select the "numpy" package by clicking on it and highlighting it.

Click the "Install Package" button at the bottom left of the "Available Packages" screen.

The "numpy" package will now load.

When the package has been installed it will return "Package 'numpy' installed successfully'" and you can close the windows to the coding screen.

Clear any code in the **TicTacToe** file and type the following code from Line 1:

```
# Creating a game of Tic-Tac-Toe using Python

# random number in Python

# import the numpy module (an array used mainly in data science)

# import the random

# import the sleep method from the time module

# (time is discussed in the next chapter & numpy arrays are explained

# in Chapter 14: FAQ)

import numpy as np

import random

from time import sleep

# This section of the code sets up the tic-tac-toe board
```

```
# numpy is a 1 to 3 dimensional array used in this game to create
# an array for the tic-tac-toe board
# an array is a list of items stored adjacent to or near each other in a sequence
# this block of code will create a blank array for the game
def new_frame():
  return (np.array([[0, 0, 0],
           [0, 0, 0],
           [0, 0, 0]]))
# this block of code checks the array for blank spaces
def blank_spaces(frame):
  sp = []
  for a in range(len(frame)):
    for b in range(len(frame)):
      if frame[a][b] == 0:
        sp.append((a, b))
  return (sp)
# Select a random place for the player
# this block of code will select the next random blank space
# available on the board for the player to make a move
def next_move(frame, gamer):
  picked = blank_spaces(frame)
  chosen_block = random.choice(picked)
```

```
    frame[chosen_block] = gamer
    return (frame)
# This section of the code determines the position
# the gamers checks are placed on the board
# determines if all the gamers check marks are in vertical order
# if they are, then the gamer wins the round
def vert(frame, gamer):
    for c in range(len(frame)):
        game_win = True
        for d in range(len(frame)):
            if frame[d][c] != gamer:
                game_win = False
                continue
        if game_win == True:
            return (game_win)
    return (game_win)
# determines if all the gamers check marks are in horizontal order
# if they are, then the gamer wins the round
def horiz(frame, gamer):
    for c in range(len(frame)):
        game_win = True
        for d in range(len(frame)):
            if frame[c, d] != gamer:
```

```
            game_win = False
            continue
        if game_win == True:
          return (game_win)
    return (game_win)
# determines if all the gamers check marks are in diagonal order
# if they are, then the gamer wins the round
def across(frame, gamer):
    game_win = True
    d = 0
    for c in range(len(frame)):
        if frame[c, c] != gamer:
          game_win = False
    if game_win:
        return game_win
    game_win = True
    if game_win:
        for c in range(len(frame)):
          yd = len(frame) - 1 - c
          if frame[c, d] != gamer:
            game_win = False
    return game_win
# This section determines who won, lost,
# or if the game was a tie
```

```
def win_check(frame):
  victor = 0
  for gamer in [1, 2]:
    if (horiz(frame, gamer) or
        vert(frame, gamer) or
        across(frame, gamer)):
      victor = gamer
  if np.all(frame != 0) and victor == 0:
    victor = -1
  return victor
# This is the section that gets the game play started
def game():
  frame, victor, turns = new_frame(), 0, 1
  print(frame)
  sleep(2)
  while victor == 0:
    for gamer in [1, 2]:
      frame = next_move(frame, gamer)
      print("Tic-tac-toe move " + str(turns))
      print(frame)
      sleep(2)
      turns += 1
      victor = win_check(frame)
      if victor != 0:
        break
```

```
    return (victor)
# This last bit of code prints out the results of the game
print(str(game()) + " Wins this round")
```

Result will print:

```
[[0 0 0]
 [0 0 0]
 [0 0 0]]
Tic-tac-toe move 1
[[0 0 0]
 [0 0 0]
 [0 0 1]]
Tic-tac-toe move 2
[[0 0 0]
 [0 0 0]
 [2 0 1]]
Tic-tac-toe move 3
[[0 0 0]
 [0 0 1]
 [2 0 1]]
Tic-tac-toe move 4
[[0 0 0]
 [2 0 1]
 [2 0 1]]
Tic-tac-toe move 5
```

[[0 0 1]

[2 0 1]

[2 0 1]]

1 Wins this round

Project 7 – Word Scramble Puzzler

```
# Word Scrambling Puzzle Game
import random
# set up the random word dictionary
def select():
    # word dictionary
    dictionary = ['window', 'donkey', 'carpet', 'hammer',
                'gutter', 'drummer', 'curtain', 'painting',
                'horse', 'dapper', 'flower', 'monkey',
                'dragon', 'python', 'tractor', 'horn']
    # system randomly picks a word
    chosen = random.choice(dictionary)
    return chosen
# create the function to randomly scramble the chosen word
def scramble(word):
    # use a method from the random module called sample
    # to scramble the randomly chosen word
    picked = random.sample(word, len(word))
    scrambled = ''.join(picked)
```

```
    return scrambled
# tally up the final score for the players
def final(ply1, ply2, pl1, pl2):
    print(ply1, 'Total score: ', pl1)
    print(ply2, 'Total score: ', pl2)
    who_won(ply1, ply2, pl1, pl2)
    print('Hope you enjoyed the game')
# find the winner of the game
def who_won(gamer1, gamer2, p1tot, p2tot):
    if p1tot > p2tot:
        print(gamer1, "won this game")
    elif p2tot > p1tot:
        print(gamer2, "won this game")
    else:
        print("The game was a tie!")
# setting up the game play function
def game_play():
    # ask gamer1 and gamer2 to enter their names
    g1name = input("Enter your name, Gamer 1: ")
    g2name = input("Enter your name, Gamer 2: ")
    # set up a score counter for each player of the game
    gp1 = 0
    gp2 = 0
    # set a counter to keep track of the number of turns
    plays = 0
```

```
# set up the loop for the game play
while True:
    word_selected = select()
    xx = scramble(word_selected)
    print("Word to unscramble: ", xx)
    # keep track of the plays to determine
    # who's turn it is
    if plays % 2 == 0:
        print('This turn is for', g1name)
        entry = input('Enter your choice: ')
        if entry == word_selected:
            gp1 += 1
            print('Score is now: ', gp1)
            plays += 1
        else:
            print('Sorry, that is incorrect.')
            print(g2name, 'You are up next!')
            entry = input('Enter your choice: ')
            if entry == word_selected:
                gp2 += 1
                print('Score is now: ', gp2)
            else:
                print('Sorry, that is incorrect! The correct word is: ', word_selected)
```

```
            again = int(input('Select 1 to play again or 0 to
end the game: '))
            # check if the gamers want to continue playing
the game or not
            if again == 0:
              final(g1name, g2name, gp1, gp2)
              break
    else:
        # Checking for gamer 2 turn
        print('This turn is for:', g2name)
        entry = input('Enter your choice: ')
        if entry == word_selected:
            gp2 += 1
            print("Your Score is :", gp2)
            plays += 1
        else:
            print('Sorry, that is incorrect!')
            print(g1name, 'You are up next!')
            ans = input('Enter your choice: ')
            if entry == word_selected:
              gp1 += 1
              print('Score is now: ', gp1)
            else:
              print('Sorry, that is incorrect! The correct
word is: ', word_selected)
```

```
            again = int(input("press 1 to continue and 0
to quit :"))
            if again == 0:
              final(g1name, g2name, gp1, gp2)
              break
        again = int(input('Select 1 to play again or 0 to
end the game: '))
        if again == 0:
          thank(g1name, g2name, gp1, gp2)
          break
if __name__ == '__main__':
  game_play()
```

Results will print:

Enter your name, Gamer 1: Tim

Enter your name, Gamer 2: Tom

Word to unscramble: rtortca

This turn is for Tim

Enter your choice: tractor

Score is now: 1

Word to unscramble: arttocr

This turn is for: tom

Enter your choice: tractor

Your Score is: 1

Select 1 to play again or 0 to end the game: 1

Word to unscramble: rdaepp

This turn is for Tim

Enter your choice: drape

Sorry, that is incorrect.

Tom, You are up next!

Enter your choice: dapper

Score is now: 2

Select 1 to play again or 0 to end the game: 0

Tim Total score: 1

Tom Total score: 2

tom won this game

Hope you enjoyed the game

Bonus Mini Project – Dice Rolling Simulator

This project is a very basic dice rolling game. There is a lot you can do with this code. As an exercise, you can do on your own, try and spruce the game up a bit by:

- Adding selection criteria
- Add players and tally scores as a dice game
- Use Tkinter to create a GUI

Create a new Python file in the Python **Projects** folder and call it **Dice**.

Clear any code in the **Dice** file and type the following code from Line 1:

```
smallest = 1
largest = 6
roll = "yes"
while roll == "yes" or roll_again == "y":
    print("Dice rolling >>>")
```

```
print("Dice has landed on >>>")
print(random.randint(smallest, largest))
print(random.randint(smallest, largest))
roll = raw_input("Roll the dices again?")
```

Result will print:

```
Dice has landed on >>>
5
2
```

Chapter 12: Working With Date and Time

Most programs will need the date and time for various reasons. When Python is installed, a date is not defined as its data type. To work with and manipulate date and time, the **datetime** module needs to be imported into the code.

The time module will also need to be imported to work with time.

The datetime module allows dates to be seen as date objects that can be used in various ways throughout the program, depending on the requirements. The time module allows the programmer to determine the system's time and work with it accurately.

To follow the examples in this section, create a new Python project called **Ch12Projects**.

Date

Create a new Python file in the Python **Ch12Projects** folder and call it **ExDate**.

Clear any code in the **ExDate** file and type the following code from Line 1 to print the current date and time:

```
import datetime

now = datetime.datetime.now()

print(now)
```

Result will print (something similar):

2020-12-25 22:35:18.559383

The above example is a very basic way of displaying and retrieving the date. It will print out the date and time in the following format:

- Date prints out the year, month, and date (yyyy-mm-dd).
- Time prints out the hour, minute, second, and microsecond (hh-mm-ss-ms)

Many methods can be used with the datetime module to return the date:

Clear any code in the **ExDate** file and type the following code from Line 1:

```
import datetime
di = datetime.datetime.now()
print(di.day)
print(di.month)
print(di.year)
```

Result will print (something similar):

25

12

2020

strftime() Method for a Date

The following code will use the **strftime()** method to pass the day of the week for today's date. It will use **datetime.now()** function to figure out what day, month, or if it is regarding the day, then return the respective string.

Clear any code in the **ExDate** file and type the following code from Line 1:

```
import datetime
di = datetime.datetime.now()
print(di.year)
print(di.strftime("%A")
```

Result will print (something similar):

2020

Saturday

The Calendar Module

Python has a module called the calendar module, which is handy for manipulating various dates for applications that require date input and output, like a diary or even a calendar.

To use the module, you will need to import the module or parts of the module when starting the code.

Clear any code in the **ExDate** file and type the following code from Line 1 to import and work with the basic calendar function:

```
import calendar
di = datetime.datetime.now()
print(di.year)
print(di.strftime("%A")
```

Result will print (something similar):

2020

Clear any code in the **ExDate** file and type the following code from Line 1 to import and work with the basic calendar function to print out a monthly calendar:

```
import calendar
cal_month = int(input("Enter the month (0 - 12): "))
cal_year = int(input("Enter the year (yyyy): "))
cal = calendar.month(cal_year, cal_month)
print("You requested a calendar for ", cal )
```

Result will print:

```
Enter the month (0 - 12): 7
Enter the year (yyyy): 2020
You requested a calendar for      July 2020
Mo Tu We Th Fr Sa Su
       1  2  3  4  5
 6  7  8  9 10 11 12
13 14 15 16 17 18 19
20 21 22 23 24 25 26
27 28 29 30 31
```

Clear any code in the **ExDate** file and type the following code from Line 1 to import and work with basic calendar function to print out a yearly calendar:

```
import calendar
year = int(input("Enter the year (yyyy): "))
cal = calendar.calendar(year)
print("Calendar for", year)
print(cal)
```

Results will print:

Enter the year (yyyy): 2020

Calendar for 2020

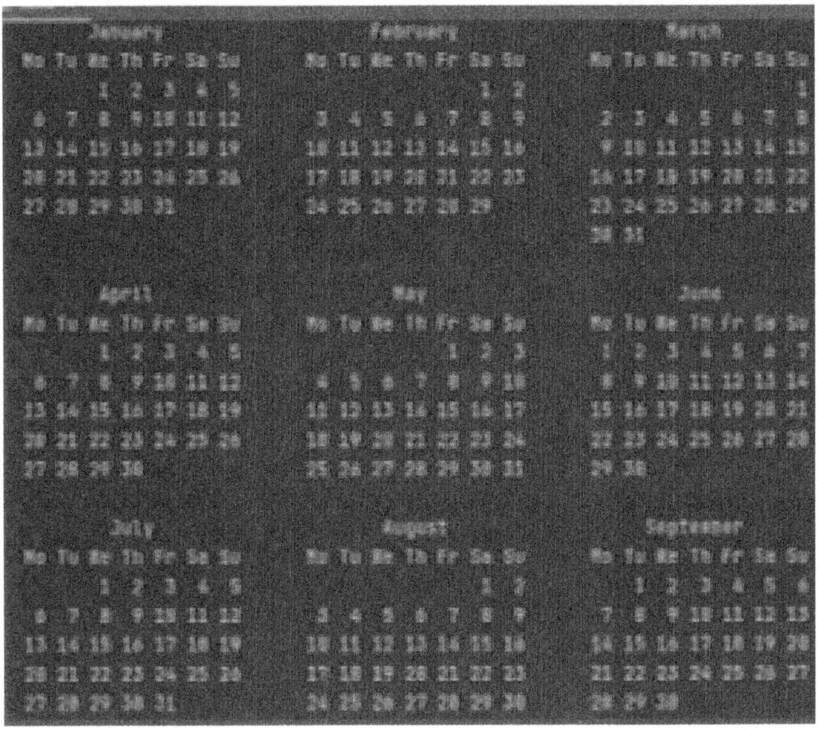

The Calendar Module

Having worked with a few calendar methods, you can see how they would be useful for working with various programs.

Syntax for the calendar() Method

The following calendar methods are some of the most widely used methods in Python:

- **calendar.weekday(yyyy, dd, mm)** – The year, day, and month fields are required for this method to work.

  import calendar

  print(calendar.weekday(2020, 7, 12))

Result will print:

6

- **calendar.month(yyyy, mm, w=0, l=0)** – The year and month are required. The width (spaces between date columns) and length (lines between date rows) are not required and can be left off altogether.

  import calendar

  print(calendar.month(2020, 7))

  Result will print:

  The calendar for the month of July will print

- **calendar.year(yyyy)** – The year is required.

  import calendar

  print(calendar.year(2020))

  Result will print:

  The calendar for the year 2020 will print

Time

There are many instances in programming where you will need to work with time. In Python, time intervals are recognized in seconds starting from **00:00:00: hrs on January 1, 1970**. This fixed point of time from which all other time is referenced is known as the **epoch**.

When you measure the passage of time in Python you are referencing it from midnight of January 1, 1970, which will always be the epoch/starting point.

Time intervals are measured in seconds as floating point values. The epoch would be 0 at midnight January 1, 1970.

Therefore midnight on January 2, 1970, would be expressed as **86 400 seconds**.

> 60 seconds in a minute, 60 minutes in an hour, and 24 hours in a day
>
> 60 * 60 * 24 = 86 400 seconds in a day

Not all systems use 1970 as an Epoch. Some systems go back a lot earlier. Earlier days than the system's epoch can still be referenced in Python; only they will have a negative value.

The **gmtime()** method from the **time** module can be used to determine the epoch of the current system you are working on.

Create a new Python file in the Python **Ch12Projects** folder and call it **ExTime** to follow along with the exercise examples in this section.

Clear any code in the **ExDate** file and type the following code from Line 1 to print the current date and time:

```
import time

print(time.gmtime(0))
```

Result will print (something similar):

```
time.struct_time(tm_year=1970, tm_mon=1, tm_mday=1, tm_hour=0, tm_min=0, tm_sec=0, tm_wday=3, tm_yday=1, tm_isdst=0)
```

The **time** module can be used to print out time in a few useful formats using various methods from the time module.

Clear any code in the **ExDate** file and type the following code from Line 1 to print the current date and time to print out the time in seconds from the system's epoch using **time()**:

```
from time import time

print(time())
```

Result will print (something similar):

1609004033.0488129

Clear any code in the **ExDate** file and type the following code from Line 1 to print the current local time using **ctime()**:

```
from time import ctime
print(ctime())
```

Result will print (something similar):

Sat Dec 26 17:35:41 2020

Time Zones

Working with time zones is for the more advanced Python user. However, it is essential to understand how they work.

Coordinated Universal Time (UTC) is the universal time zone standard. It is the time zone that all other time-keeping clocks around the globe are synchronized. UTC uses the astronomical time to measure its time. Astronomical time is set on the Earth's rotation.

Using UTC as the baseline, time zones are then offset by their geographical location, for example:

- Central Time Zone (CT) = 5 or 6 hours behind UTC.

  UTC - 6:00 or UTC - 5:00

Time zone calculation can get confusing and complicated as some countries/cities also use daylight savings time, which goes back by a particular time in the summer months.

Time Ticks

The **time.time()** function in Python will return the current time in ticks since the epoch to either a specific point in time or the system's local current time. A tick is much like that seconds hand on a clock that ticks off the date's seconds.

Time Tuple

Python time does not always need to be represented by a number. It can also be defined as a tuple. Time tuples can extract certain elements of time and return them in a more readable format.

```
import time
timet = (2019, 3, 10, 8, 50, 6, 6, 69, 1)
print(time.mktime(timet))
```

Result will print:

```
1552204206.0
```

The above result will round down the floating pointing value, returning a whole number.

Time Formatting

There are a few ways to convert time in Python to a more readable format other than use tuples.

The example below uses asctime() from the time module in Python, which is used to convert time tuples to a timestamp. Each of the printed time statements returns date and time in a more structured and readable format.

```
import time
time.asctime(time.gmtime())
print(time.asctime(time.gmtime()))
print(time.asctime(time.localtime()))
```

Result will print:

```
Sat Dec 26 19:30:34 2020
Sat Dec 26 19:30:34 2020
Sat Dec 26 19:30:34 2020
```

Strftime() Method Data and Time Codes

The following table is a list of the legal format codes that can be used with the **strftime()** method in Python.

It is essential to note the case when using the format codes below because an uppercase and lower case character return different outputs.

Code	Description		Output
%a	Returns the weekday	Short version	Mon
%A	Returns the weekday	Full version	Monday
%w	Returns the weekday	Shows as a number	7
%d	Returns the day of the month	Shows as a number	01
%b	Returns the month	Short version	Jan
%B	Returns the month	Full version	January
%m	Returns the month	Shows as a number	12

%y	Returns the year	Shortened version - excludes the century	20
%Y	Returns the year	Full version	2020
%H	Returns the hour — 24-hour clock	Shows only the hour (00 - 23)	23
%I	Returns the hour — 12-hour clock	Shows only the hour (00 - 12)	12
%p	Returns AM or PM	Shows as AM or PM	AM
%M	Returns the minute	Shows only the minute (00 - 59)	41
%S	Returns the second	It shows only the second (00 - 59)	10
%f	Returns the microsecond	Shows only the microsecond (0 - 999999)	878795
%z	Returns the UTC timezone offset — Coordinated	+0100	

	Universal Time		
%Z	Returns the Timezone – Central Standard Time	CST	
%j	Returns the day of the year as a number	Shows as a number (1 - 365)	225
%U	Returns the week of the year – If the first day of the week is Sunday	Shows as a number (00 - 52)	31
%W	Returns the week of the year – If the first day of the week is Monday	Shows as a number (00 - 52)	32
%c	Returns the local version of the date and time	Shows the format in the way the systems OS is setup	Wed Jan 1 2021:18:00:00
%x	Returns the local version of	Shows the format in the way the	30/12/2020

	the date	systems OS is setup	
%X	Returns the local version of the date and time	Shows the format in the way the systems OS is setup	12:00:05
%%	Returns a % character		%

Chapter 13: Work With Date and Time Practice Projects

The best way to learn about working with date and time in Python is to do a few projects. This chapter will test your current knowledge of Python and give you insight into some more advanced Python programming methods.

Project 8 — Calculate My Age

This project uses a date input to determine a person's age.

Create a new Python file in the Python **Projects** folder and call it **AgeCalc**.

Clear any code in the **AgeCalc** file and type the following code from Line 1:

```
# Calculate my age

# import the datetime module

import datetime

# ask the user to enter their birthdate

dob = input("Please enter your birthdate as dd/mm/yyyy: ")

dob = datetime.datetime.strptime(dob, "%d/%m/%Y").date()

print("As your birthday is on the "+ dob.strftime("%d") + " of " + dob.strftime("%B, %Y"))
```

currentDate = datetime.datetime.today().date()

 # calculate the difference between the current year and year of birth

age = currentDate.year - dob.year

monthcalc = currentDate.month - dob.month

datecalc = currentDate.day - dob.day

 # a type conversion is needed to convert a string to an integer

age = int(age)

monthcalc = int(monthcalc)

datecalc = int(datecalc)

 # calculate age

if monthcalc < 0 :

age = age-1

elif datecalc < 0 and monthcalc == 0:

age = age-1

 # print the out of what the users age is

print("Your current age is {0:d}".format(age))

Result will print:

Please enter your birthdate as dd/mm/yyyy: 05/04/1995

As your birthday is on the 05 of April, 1995

Your current age is 25

Project 9— Is it a Leap Year?

This project uses a date input to determine if the year is a leap year or not.

Create a new Python file in the Python **Projects** folder and call it **LeapYear**.

Clear any code in the **LeapYear** file and type the following code from Line 1:

```
# Python program to check if a year is a leap year or not
# ask user to enter the year
check_year = int(input("Please enter the year to check: "))
# check for leap year
if check_year % 4 == 0 and check_year % 100 != 0:
  print("The year", check_year, " is a Leap Year")
elif check_year % 100 == 0:
  print("The year", check_year," is not a Leap Year")
elif check_year % 400 ==0:
  print("The year", check_year," is a Leap Year")
else:
  print("The year", check_year," is not a Leap Year")
```

Result will print:

Please enter the year to check: 2007

The year 2007 is not a Leap Year

Project 10 — Date vs. Day of the Week

This project uses a date input to determine the day of the week for a given date using Python's calendar module.

Create a new Python file in the Python **Projects** folder and call it **DayofWeek.**

Clear any code in the **DayofWeek** file and type the following code from Line 1:

```
# Check what day a date falls on
# import the datetime and calendar module
import calendar
# create a function to find what day of the week the entered date falls on
def getday(dow):
  day, month, year = (int(i) for i in dow.split(' '))
  dayno = calendar.weekday(year, month, day)
  weekdays = ["Monday", "Tuesday", "Wednesday", "Thursday",
        "Friday", "Saturday", "Sunday"]
  return (weekdays[dayno])
# as the user to input the query date
dow = input("To find the day of the week, enter the date as dd mm yyyy: ")
print(getday(dow))
```

Result will print:

To find the day of the week, enter the date as dd mm yyyy: 03 02 2003

Monday

Project 11 – Clock

This project will demonstrate how to create a simple clock application in Python using the datetime, time, and winsound modules. The winsound module is a module that contains various sounds that a programmer can prompt code to play. For instance, the beeping of an alarm clock, as demonstrated below.

This project uses a few simple Python intermediate level instructions, like working with the TK GUI interface methods, and a few more.

Create a new Python file in the Python **Projects** folder and call it **Clock1**.

Clear any code in the **Clock1** file and type the following code from Line 1:

```
# A GUI interface clock timer

# import the modules that will be needed for the program

# the Tkinter module has the files needed to create the GUI interface for the clock

# the winsound module holds the sounds required for the alarm clock

import winsound

import datetime

import time

from tkinter import *

# create a functional while loop to set the alarm timer

def clock_alarm(timer_set_up):

   # while the current condition is
```

```
 while True:
     # the alarm will sleep until the user sets a time
     time.sleep(1)
     # find out what the current time and date is
     time_is_now = datetime.datetime.now()
     time_now = time_is_now.strftime("%H:%M:%S")
     date_now = time_is_now.strftime("%d/%m/%Y")
     # print out the currently set date and time
     print("The date is set to:", date_now)
     print(time_now)
     # determine if the current time equals the set
alarm
     if time_now == timer_set_up:
         print("It is now time to wake up")
         # use the beep sound as the alarm
         frequency = 440
         duration = 200
         for i in range(0, 9):
             winsound.Beep(frequency, duration)
             frequency += 440
             duration += 200
         break
# sets the actual time on the clock
def clock_time():
```

```
    timer_set_up =
f"{hour.get()}:{min.get()}:{sec.get()}"
    clock_alarm(timer_set_up)
# creating the clock GUI
timeclock = Tk()
timeclock.title("My Alarm Clock")
timeclock.geometry("400x200")
settime_in = Label(timeclock, text="24 Hour Format",
fg="blue", bg="white",font="Georgia").place(x=60,
y=120)
settime = Label(timeclock, text="Hour:   Min:   Sec:   ",
font=60).place(x=110)
set_alarm = Label(timeclock, text="Wake up at: ",
fg="blue", relief="solid",
                font=("Georgia", 7, "bold")).place(x=0, y=29)
# initialize the alarm
hour = StringVar()
min = StringVar()
sec = StringVar()
# enter the time
enter_hour = Entry(timeclock, textvariable=hour,
bg="pink", width=15).place(x=110, y=30)
enter_minute = Entry(timeclock, textvariable=min,
bg="pink", width=15).place(x=150, y=30)
enter_second = Entry(timeclock, textvariable=sec,
bg="pink", width=15).place(x=200, y=30)
# To take the time input by user:
```

```
accept_time = Button(timeclock, text="Set Alarm",
fg="red", width=10,
command=clock_time).place(x=110, y=70)

timeclock.mainloop()
```

Results will print:

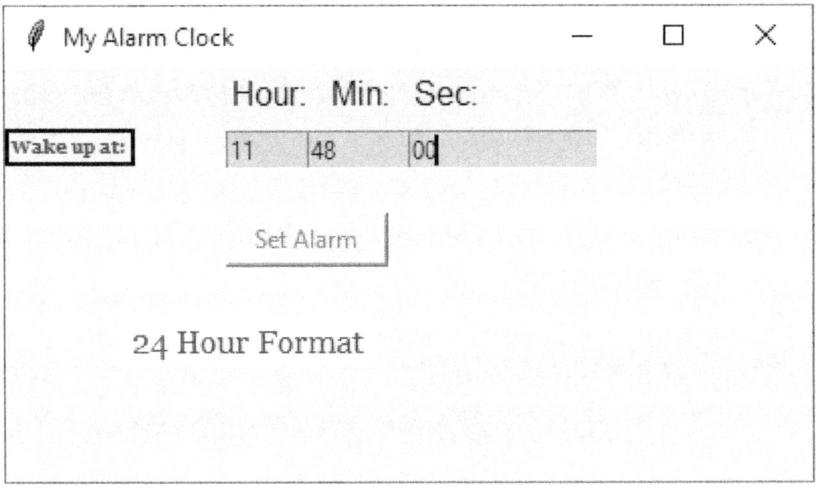

The date is set to: 26/12/2020

11:48:00

It is now time to wake up

Project 12 - Calendar and Date Picker

This project will demonstrate how to create a calendar by picking a year in Python using the calendar module in Python.

This project uses a few simple Python intermediate level instructions, including using the TK GUI interface methods.

Create a new Python file in the Python **Projects** folder and call it **DatePickCal**.

Clear any code in the **DatePickCal** file and type the following code from Line 1:

```
# Yearly Calendar
# import the calendar module
import calendar
# import all the methods from tkinter (the GUI interface program)
from tkinter import *
# create the calendar format
def yearcal():
    # set out the look of the calendar with tkinter
    cal_frame = Tk()
    cal_frame.title("YEARLY CALENDAR")
    cal_frame.geometry("550x600")
    cal_frame.config(bg="White")
    # convert the current text
    find_year = int(yr_input.get())
    # find the correct calendar for the selected year
    cal_items = calendar.calendar(find_year)
    # create a variable to store the contents of the calendar
    yearly_cal = Label(cal_frame, text=cal_items, font="Consolas 10 bold")
    # create a grid to display the calendar
    yearly_cal.grid(row=5, column=1, padx=20)
    # the GUI is ready to be created
```

```
    cal_frame.mainloop()
# main code
# creates the gui, sets up the features for the gui
# like the bg (background color, font, gui screen size, etc.
if __name__ == "__main__":
    screen = Tk()
    screen.title("YEARLY CALENDAR")
    screen.config(bg="White")
    screen.geometry("375x250")
    yc = Label(screen, text="YEARLY CALENDAR", fg="Dark blue", bg="Blue",
               font=("Helvetica", 28, 'bold'))
    # label for the date entry
    yr = Label(screen, text="Year: ", fg="Dark blue", bg="Light blue")
    # year input field
    yr_input = Entry(screen)
    # a call function attached to a submit button to submit the year
    Select = Button(screen, text="Get Calendar", fg="Dark blue",
                    bg="Light green", command=yearcal)
    # Create an Exit Button and attached it to the exit function
    Exit = Button(screen, text="Exit", fg="Black", bg="Red", command=exit)
```

```
# create a grid to display each field for the calendar
yc.grid(row=1, column=1)
yr.grid(row=2, column=1)
yr_input.grid(row=3, column=1)
Select.grid(row=4, column=1)
Exit.grid(row=6, column=1)
# run the GUI
screen.mainloop()
```

Result will print:

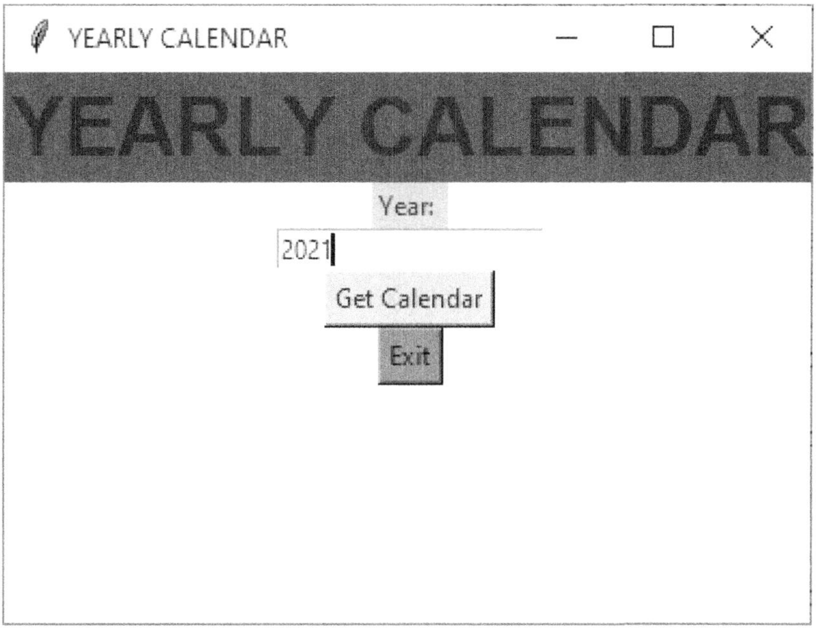

Type in a year, and it will print out the calendar for the year, keep in mind this is a very basic date picker for this course's scope.

Chapter 14: Python FAQ

Here are some of the top frequently asked questions for Python.

Question 1: What are Python's key features?

Python is an interpreted language; unlike other languages such as C++, etc., it does not need to be compiled before the program can be executed.

Python allows for classes to be defined with inheritance and composition, making it suitable for object-orientated programming.

Python classes are first-class objects — they can be assigned, passed into functions, and returned from functions.

Python functions are first-class objects — they can be assigned, passed into functions, and returned from functions.

Types of variables do not need to be stated as Python is dynamically typed.

Python can be used for data science applications, big data analysis, business applications, web development, gaming applications, and more.

Python is a simple programming language to learn.

Question 2: What is an interpreted language?

An interpreted language is a programming language that is run through an interpreter that translates programming code to a language the machine processor understands.

Question 3: What is the difference between a compiler and an interpreter?

Both compilers and interpreters are used to convert code into a language that the computer can understand to execute the program.

Compilers and interpreters are needed because programs are written in a language that the programmer can understand. This is known as a high-level language (source code), but it is not a language the computer understands.

Once the program is ready to be executed, the source code must be converted into machine code, which is where the compiler or interpreter comes in.

Compilers will take the code and create the program. While they are converting the source code, the compiler will run through it, checking for errors. If the source code comes out error-free, the compiler will turn the source code into machine code.

Interpreters execute the source code line by line as the program is run. They do not generate machine code, nor do they link files as a compiler does.

Question 4: What are Python modules?

Python modules are files that contain Python definitions and statements. The file is recognized with a .py extension and is a file that contains Python code.

Modules break down large programs into smaller, more manageable ones.

Question 5: What are the most important Python modules?

datetime – Offers different ways to represent date and time in Python code.

Django – Important for web development to create manageable and secure websites.

JSON – This module provides API support for converting in-memory Python objects to JavaScript Object Notation (JSON).

math – This module provides more mathematical function support for Python.

os – This is Python's standard utility module that provides functions that can interact with the operating system.

random – This module offers Python the ability to generate random numbers.

sys – This module has variables and functions that are used to manipulate the Python runtime environment.

Question 6: What are the Python local variables?

A local variable is one that can only be read in its local environment. For instance, a variable that has been defined within a function cannot be used or recognized outside the scope of that function.

Question 7: What are the Python global variables?

A global variable can be read within the scope of the function it was defined in and outside that function's scope.

Question 8: What is Python type conversion?

Converting one data type into another in Python is known as type conversion.

Question 9: What are the Python type conversion functions?

int() – Conversion of all data types into an integer type.

float() – Conversion of all data types into a float type.

ord() – Conversion of characters into an integer.

hex() – Conversion of integers to hexadecimal.

oct() – Conversion of an integer to octal.

tuple() – Converts data types to a tuple.

set() – After converting to set, this function returns the data type.

list() – Converts any data type to a list type.

dict() – Converts an ordered tuple into a dictionary type.

str() –Converts an integer into a string.

Question 10: What is the difference between a list and an array in Python?

Both a list and array store data. They can both store integers and strings. They can also both be iterated as well as indexed.

Arithmetic functions can be performed on an array. Python will return an error if an arithmetic function is performed on a list.

Arrays need to be declared as they are not part of Python's built-in functions. Lists do not need to be declared as they are part of Python's built-in functions.

Array stores data more compactly and efficiently compared to lists. Thus an array is more suitable for storing large amounts of data.

Question 11: What are iterators in Python?

An iterator is an object capable of returning turn data one element at a time and can be traversed through and iterated upon.

Question 12: What is a Python docstring?

A docstring is usually the first statement that appears in a class, function, module, or method, and it is a string literal. It is used to comment/document on a block of code and explains what the code does, unlike code comments that would explain the how or why of lines or blocks of code.

Question 13: What are the main Python libraries?

Scrapy – This is used for data mining.

BeautifulSoup – This is used for data mining.

NumPy – This is used for data processing and modeling.

PyTorch – This is used for data processing and modeling.

SciKit-Learn – This is used for data processing and modeling.

Pandas – This is used for data processing and modeling.

TensorFlow – This is used for data processing and modeling.

Matplotlib – This is used for data visualization.

Plotly – This is used for data visualization.

Bokeh – This is used for data visualization.

pydot – This is used for data visualization.

Question 14: What is Django?

Django is a free open source high-level Python Web framework for web development.

Question 15: What is a Python lambda function?

A lambda function is a function that is defined without a name.

For example, a normal function is defined with a user-defined name:

def my_function()

A lambda function runs anonymously without a name.

*calculate = lambda a: a * 5*

The above lambda function would be similar to the following normal function definition:

```
def calculate(a):
    return a * 5
```

Question 16: What is a NumPy array?

A NumPy array is a much faster, more efficient, and less memory hungry type of list function. NumPy is one of Python's top data science modules because of its vast array of capabilities and convenient data manipulation methods.

A Numpy array is expressed as a grid of the same type of values that are tuple indexed.

Question 17: What is Concatenation?

While programming, you may have two strings of data that need to be joined together to make a new string. This is known as concatenation, and the new string that will be used to contain both string values is known as the string object.

For example:

a = "Year is: "

b = "2020"

print(a + b) # this section of the code joins a string to a string using the print function.

Result will print:

The year is: 2020

Question 18: What is the Best Way to Debug a Python Program?

The Python pdb module has all the methods and tools needed to debug a Python program.

It can be run from within a script by importing the pdb module and using pdb methods on various script parts.

Python pdb module can also be used at the command line prompt. This debugging method requires the programmer to know what the Python program's file name and the directory location it is stored under.

Question 19: What is Python Multithreading?

Python does not support multithreading and uses something called the Python Global Interpreter. This interpreter will limit a multiprocessor machine to running only one thread at a time by placing Lock limits.

Multithreading is the process of running multiple concurrent threads simultaneously.

Question 20: What are *args and **kwargs?

When a function is set up with a certain number of parameters, Python assumes that the same amount of arguments will be fed to those parameters. If the program is a small one, it is easy to pass an equal number of arguments to a statement's parameters.

There are plenty of applications where the number of arguments being passed to a function is unknown. A programmer can use *args when they are uncertain of how many arguments will be passed through to a statement.

Likewise, **kwargs is used to define an unknown number of keyword arguments required for a statement.

Conclusion

Currently, Python is one of the most sought after programming skills in IT. As Python goes from strength to strength with every new release, it will be a top sought after skill for years to come.

One of Python's biggest growing markets is in the data science field, which incorporates machine learning, predictive analysis, and artificial intelligence. As you get more advanced with Python programming, you will learn about the thousands of libraries available to Python. Python libraries include Pandas, Numpy, PyTorch, and Tensorflow. These programs are fast becoming the go-to Python libraries for the analysis of big data.

Python is also a valuable tool for web development, business applications, robotics, and gaming. Now that you have completed *Python For Beginners: A crash course to learn Python Programming in 1 Week,* you are one step closer to moving into your chosen programming field.

Keep practicing your new-found Python skills, and soon you will be ready to advance to the intermediate level of Python programming.

Good luck and keep coding!

Other Books from the Author

Python For Beginners:
A crash course to learn Python Programming in 1 Week

Python Machine Learning:
A Step by Step Beginner's Guide to Learn Machine Learning Using Python

Python Data Analysis:
A Beginners Guide to Master the Fundamentals of Data Science and Data Analysis by Using Pandas, Numpy and Ipython

References

Data Types. (n.d.). Python.org. https://docs.python.org/3/library/datatypes.html

Difference between Compiler and Interpreter. (2019, May 27). Business Insider. https://www.businessinsider.in/difference-between-compiler-and-interpreter/articleshow/69523408.cms#:~:text=Interpreter%20translates%20just%20one%20statement,the%20process%20is%20much%20slower.

History of Python. (2019, May 6). GeeksforGeeks. https://www.geeksforgeeks.org/history-of-python/

How to Think Like a Computer Scientist: Learning with Python 2nd Edition documentation. (n.d.). Open Book Project. https://www.openbookproject.net/thinkcs/python/english2e/ch02.html#:~:text=A%20statement%20is%20an%20instruction,result%2C%20if%20there%20is%20one.

Kite. (2020, March 11). *How to build HANGMAN with Python in 10 MINUTES* [Video]. YouTube. https://www.youtube.com/watch?v=m4nEnsavl6w

Print a Calendar in Python. (n.d.). Tutorialspoint. https://www.tutorialspoint.com/print-a-calendar-in-python

Print Function and Strings. (n.d.). Python Programming.net. https://pythonprogramming.net/python-tutorial-print-function-strings/

Python — Date & Time. (n.d.). tutorialspoint. https://www.tutorialspoint.com/python/python_date_time.htm

Python For Loops. (n.d.). w3schools.com. https://www.w3schools.com/python/python_for_loops.asp

Python If ... Else. (n.d.). w3schools.com. https://www.w3schools.com/python/python_conditions.asp

Python Iterators. (n.d.). Programiz. https://www.programiz.com/python-programming/iterator

Python PIP. (n.d.). w3schools.com. https://www.w3schools.com/python/python_pip.asp

Python Program to implement jumbled word games. (2019, July 19). GeeksforGeeks. ttps://www.geeksforgeeks.org/python-program-to-implement-jumbled-word-game/?ref=lbph

Python program to implement Rock Paper Scissor game. (2020, October 23). GeeksforGeeks. https://www.geeksforgeeks.org/python-program-implement-rock-paper-scissor-game/

Python Program to Make a Simple Calculator. (n.d.). Programiz. https://www.programiz.com/python-programming/examples/calculator

Python strftime(). (n.d.). Programiz. https://www.programiz.com/python-programming/datetime/strftime

Python Strings. (n.d.). Google for Education: Python. https://developers.google.com/edu/python/strings#:~:text=Python%20has%20a%20built%2Din,quotes%20are%20more%20commonly%20used.

Python TextCalendar Module : month() method. (2020, February 28). w3resource. https://www.w3resource.com/python/module/calendar/month.php

Python While Loops. (n.d.). w3schools.com. https://www.w3schools.com/python/python_while_loops.asp

Python Virtual Environment | Introduction. (2020, November 27). GeeksforGeeks. https://www.geeksforgeeks.org/python-virtual-environment/

Ronquillo, A. (n.d.). *A Beginner's Guide to the Python time Module*. Real Python. https://realpython.com/python-time-module/#the-epoch

Simpao, C. (2020, September 21). *10 Best Python IDE & Code Editors (Updated Guide)*. hack.io. https://hackr.io/blog/best-python-ide

Suarez, A. (2020, August 6). *Alarm Clock Using Python With Source Code.* https://itsourcecode.com/free-projects/python-projects/alarm-clock-using-python-with-source-code/

What is the difference between single and double quotes in Python? (n.d.). Tutorialspoint. https://www.tutorialspoint.com/What-is-the-difference-between-single-and-double-quotes-in-python#:~:text=A%20string%20as%20a%20sequence,representations%20can%20be%20used%20interchangeably.

Ufeng, G. (2017, November 9). *Which Python package manager should you use?* Towards data science. https://towardsdatascience.com/which-python-package-manager-should-you-use-d0fd0789a250#:~:text=Anaconda,-Now%2C%20if&text=Anaconda%20is%20popular%20because%20it,isolate%20different%20libraries%20and%20versions.

Printed in Great Britain
by Amazon

49937615R00129